Most lovingly dedicated to my wife, Consolacion (Conching/Connie/Mamanie); our children Michele/Roland, Christian/Maricel and Jude/Arlyn; our grandchildren Christine/Hanzel, Rochele, Miro, Charles, Angel and Blessie; and our great grandchildren Cianna and Drake.

Many thanks to Percival (Percy) Campoamor Cruz for inspiring and egging me to complete my life story which has been in my personal "cloud" for about 16 years.

BUILDING MONUMENTS IN THE SKY

(*Filling the World with Love*)

By

Bernardo M. Soriano, Jr.

The world may be full of warts and all, but it's still a wonderful world! Thank you, God! (BMS).

In the morning of my life ...

The world was in turmoil when I was born at the Philippine General Hospital (PGH) in Ermita, Manila on 17 May 1943. World War II was going on! The allied military forces of the USA, Great Britain, the Soviet Union and China were battling Germany and Italy in Europe and Japan has invaded the Philippines. Manila was being sieged and bombed when I was about a year old. My mother told me that I was crawling and still learning how to walk when a bomb explosion near our place in Singalong made me suddenly stand up and run towards her. My father was a Philippine guerilla, who sustained a rifle wound from a sniper perched on a tree. The bullet entered his inner right thigh and exited, leaving a large wound on his outer right thigh. He was confined at the Philippine Orthopedic Hospital in Mandaluyong, Rizal for several weeks. My mother would lug me and carry my sister and walk to Sta. Ana to take a ferry banca across the Pasig River to visit my father, who I remembered seeing with a plaster cast on his hanging leg. These situations turned out to be the general patterns throughout my life: hardship, struggle, success and subsidence.

My Family

My mother, *Angela M. Soriano*, who we called Mama Aba (Aba is short for Taba, in contrast to her mother who was called Nanay Payat), was the second among the 13 children of *Inocencio Monillas Sr.* and *Constancia de Guzman*, who were both originally from Bangar, La Union. Her siblings were: Dionisio (Diony), Nieves (Nieveng), Amanda (Manding), Celso, Luz, Romeo (Butog), Paulina (Mimi), Alberto (Berting),

Jesusa (Chu-chu), Raul, Inocencio (Nonoy/Inteng) Jr. and Constancia (Connie). Tatay, as we called my grandfather, was the 6th certified public accountant of the Philippines and he was the auditor of the Philippine Veterans Board when I was a boy. We called my grandmother Nanay, who sold jewelries she got from Bambang, Sta. Cruz, Manila. Being the oldest of their 44 grandchildren, Nanay pampered me by giving me a necklace and a ring. She was a sweet and gentle woman whose favorite song was "Blowing Bubbles" which she sang when she was happy, especially after finishing half of a "lapad" (flat bottle) of Manila Rum. She gulped the liquor directly from the bottle and shared it with my mother for appetite before lunch, as they said. They drank the other half before dinner. Mama and my other relatives called me "Bolonging" because of Mama's favorite mahjong tile, which is one ball. She told me that when she was pregnant with me while playing the game and, she had the tile as her "puro" (last piece to win), she would say while feeling it with her middle finger "C'mon, one bolonging"! This nickname became a real thing when a boil grew on my right temple that left a circular scar about an inch in diameter when I was a boy. Later on, when I was already a teenager, my pet name became "Ging" which stuck with me until my adulthood. I consider now those who still call me "Bolonging" or "Ging" as people who are close or endeared to me.

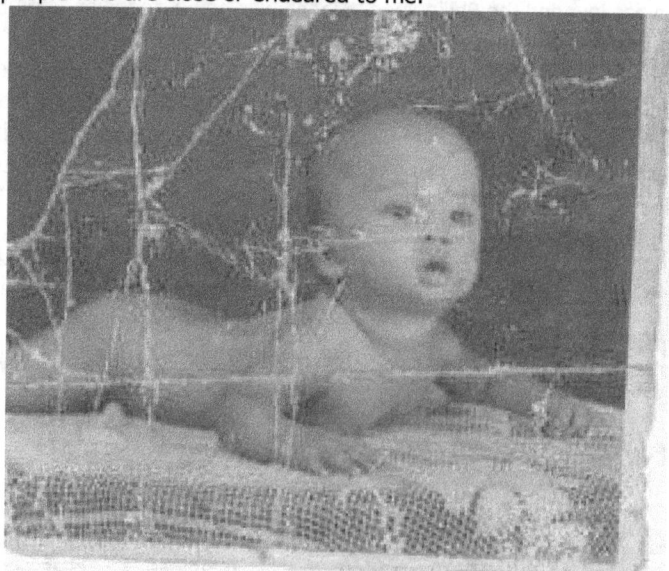

3-month old baby Bolonging

7-year old feverish Ging on Echague St., Quiapo, Manila.

My father, *Bernardo* (*Nanding*) *C. Soriano Sr.*, who we called Daddy, was a taxi driver while my mother was a full-time housewife. He was the second of the eight children of *Juan Soriano* and *Pilar Cruz*, who were originally from Binmaley, Pangasinan and Solano, Nueva Vizcaya, respectively. Lolo Juan, called Papang, was already gone when I was born. I was told later that he worked as an internal revenue collector. Lola Pilar, who we called Mamang, finished only the sixth grade and was a former teacher in the elementary school. She was an industrious woman, who worked the house almost all day. She had a "green thumb", which accounted for the beautiful plants around their small house on G. del Pilar St., just a few houses away from our house on Taal St. in Singalong, Manila. Mamang and her children spoke fluent Spanish. I learned later that my paternal great grandfather was a pure-blooded Spaniard, who was a disciplinarian.

Daddy solved the crossword puzzles of the Manila Times while he was in the taxi line at the Manila Hotel in Luneta. He was so good at it that he completed the puzzle in less than 20 minutes. This influenced me later to be my favorite pastime aside from sudoku. He had a good penmanship, which I barely got.

We lived in the Monillas's house, which is a two-floor structure on Dagonoy (later named Florentino Torres) and Taal streets. The house was built in 1925 and was typical of the houses at that time, with large window panes made of capiz shells and small lower windows and galvanized iron roofs. The frames and rafters of the house were made of hard wood that were resistant to termites. It endured the ravages of the war, and typhoons, floods and earthquakes that occurred frequently in the country. It was torn down in the late 2000's when the lot on which it stood was sold by uncle Raul, who was the designated administrator of the property when my grandmother died at almost 101 in Seattle, Washington, USA in 2009. My grandfather passed away in 1951 after a heart attack while playing mahjong.

The Monillas siblings: (L-R) Diony, Angie, Nieveng, Manding, Celso, Luz, Butog, Mimi and Berting. Not in photo are: Chu-chu, Raul, Nonoy and Connie.

We occupied a small room on the first floor, which we shared with my uncle Celso and his family of five, and lolo Cosme and his family of three. Uncle Celso lived with his wife, auntie Tessie, and their children. Lolo Cosme, Nanay's younger brother, lived with his wife, Lola Bruna and children David and Boy. We had a common large dining table between our rooms and toilet/bath rooms. The kitchen was outside the house. We used fire woods, called bakawan, as our fuel for cooking. A large camachile tree stood at the back of the house from which I occasionally plucked its fruits with a long stick and a wire hook at the end. There was once a time when this strange, dark man with a wide jaw and slightly bearded face, who was barefoot and half-naked, suddenly appeared and offered to get the fruits for me. He climbed the tree like a cougar, crouching on a nearly horizontal thorny branch without difficulty or injury.

He disappeared as suddenly as he appeared after giving me the camachile fruits. He was the same man who earlier intervened between me and a neighborhood bully when we were about to square off to a fist fight in the street. He disappeared as quickly as he appeared after breaking up the fight. I thought then that he was my guardian angel.

I have a vague knowledge of the personal background of my maternal relatives. I was the ring bearer when Uncle Diony married Auntie Biring, nee *Barbara Manahan*, at the St. Anthony Church in Singalong, Manila. They had eight children, namely; Marietta, Reynaldo, Wilfredo, Dionisio Jr., Gilberto, Efren, Merceditas and Francisco. He became a pianist, initially worked at the Bayside Nightclub along Roxas Boulevard, then in a cruise liner, until he suffered a stroke. Auntie Nieves, who was my baptismal godmother, was a lawyer and married Atty. *Rosendo Alcalde* with whom they had four children. They are Maria Socorro, Maria Consuelo, Jose Emilio and Maria Rosario. Auntie Amanda was married to *Julio Mijares*, with whom she had four children, namely; Samuel, Daniel, Mildred and Michael. She worked at the Philippine Constabulary SOCIA (Supervisory Office for Security and Investigation Agency) until she retired. Uncle Celso was an architect, who graduated from the Mapua Institute of Technology (MIT) in Sta. Cruz, Manila. I vividly remember the time he usually comes home from MIT at about 1 p.m. when the theme song of radio station DZMT played "Summer Love". Seeing the willow trees now in Canada brings back memories of this time. He married our beautiful, mestiza neighbor *Teresita Lero*, with whom he had five children, namely; Angelita, Maria Elena, Edgar, Elmer Edsel and Ellinor. They all live now in Seattle, Washington, USA, except Angelita who died after birth. Auntie Luz finished college at St. Paul in Manila, was a businesswoman and married *Potenciano Mendoza*, with whom she had a daughter named Marylou (Malu). She and Malu are now in San Jose, California, USA. Uncle Butog (also, Butch) was a good singer, who joined *Pepe Pimentel* and *Bert Nievera* in the radio program "Melody in the Air" in the 1950s. He married *Rosa Humangit* from Hinunangan, Samar and had six children, namely; Rosanna, Roberto, Rowena, Rolando, Romeo Jr., and Rogelio. They moved to Seattle, WA, except Rosanna who settled in Surrey, Canada with her own family. Uncle Butch lost his voice after some years in Seattle and used a microphone to enhance his voice from his voice box. Auntie Mimi migrated to the USA first, and completed a nursing course to enable her to work as such in a hospital in Seattle. She got married to a man there with the family name Edwards, with whom she bore a daughter, she named Evelyn. Uncle Berting was married to *Sonia Murata* and had three children with her. Their children are Maria Teresa, Alberto Jr. and Mary Ann. They separated and Uncle Berting lived with *Myrna Jorge* who bore him a daughter named Miraflor. He was sponsored by his elder sister Mimi to immigrate to the USA, leaving Myrna and her daughter behind. He eventually got Miraflor to the USA. Auntie Chuchu married *Proceso (Ging-ging) Anunciado* and bore a son, they named Manolo. He left his family for reasons I have no knowledge of. Auntie Chuchu and Manolo were petitioned by Auntie Mimi to join her in Seattle, where she met and married *Danilo Victorio*. They had a daughter, named Jilli. Uncle Raul was also petitioned by Auntie Mimi to live in Seattle. He worked as a cook in a fishing boat that sailed to Alaska. He met and married *Robin Rutledge* with whom he had two children, Antony and Delora (Lora). Uncle Nonoy was also petitioned by Auntie Mimi to Seattle. He married *Normita Encomienda* and had three children they named Ian, Norman and Monica. Auntie Connie did not marry until she got to work in the USA as a practicing nurse. She married *Joseph Byrne* and still is childless. They live in Alexandria, Virginia, USA. I am proud to say that I am the oldest while Lora is the youngest of the 44 Monillas cousins.

I learned in Taal St. that one's economic status adversely or favorably affects everyone, including one's close relatives. There was a time when one of my uncles, who was really nice and kind, yelled invectives at us during his alcohol intoxication, telling us that we are not welcome in the house of his parents. I cannot forget my confirmation godfather, *Daniel Mijares*, who lived on Fuentes St., next to Taal St. While having beer with my father outside our house, he told me to leave the place for my own sake. As I was already working then, he advised me to "love your work". I kept his advice in mind throughout my career.

Photo above shows Bernardo and Angela Soriano with Ging, Eddie and Beth in 1946 (Eddie is partially visible at right between Mama Aba and Daddy).

Photo above is Beth's baptism on Sep. 9, 1946 with ninang Miring and ninong *Winnie Calalang* seated and our parents standing with our grandmothers between them. To the left of Mama Aba is lola Nieves, auntie Miring's mother, and Nanay. I am standing beside uncle Winnie. Lola Nieves and Nanay are the spouses of brothers Benigno and Inocencio Monillas, respectively.

Daddy's siblings were Socorro (Corring/Inday), Luz (Lucing), Juliet, Feliciana (Nanette), Antonio (Tony), Juanita (Nita) and Lourdes (Lulu). Shown in the photo below are his parents and sisters Corring (standing at right), Lucing and him (sitting). He first worked for a neighbor's taxi fleet when I was a boy, then for the Golden Taxi until he retired in the late 1960s. He was a hardworking and honest man but sometimes bad luck haunted him. I remembered seeing him come home one day with blood on his taxi uniform. He told me that he had a fight with a fellow driver at the garage on Canonigo St., Paco, Manila and his experience as an amateur boxer came in handy in defending against a bigger man (Daddy once related to me that he was an amateur boxer at the Far Eastern University (FEU) and he often sparred with Speedy Cabanela, a.k.a. *Materno Lumacad*, a national lightweight champion in 1952). He bloodied his nose and got the blood splattered on his uniform. At another time, he came home with blood on his face from a driving accident. He narrated that while overtaking a cart being pulled by a carabao along the highway going to the south, he miscalculated his maneuver and had to swerve to his right to avoid a head-on collision with another vehicle. In doing so, his taxi which was the earlier version of the Minica, turned turtle and plunged to the rice field. Luckily, he did not suffer any serious injury except for a small cut on his eyebrow.

My knowledge of my paternal relatives is likewise quite limited. I know that Auntie Inday was a doctor of medicine and worked as the city health officer in Jaro, Iloilo when I lived with her and her family in 1960-1961. She was married to *Antonio Valdez*, who was a judge in Ivisan, Capiz. They had five children, namely; Antonio Jr. (Jun), Sandra (Sandy), Sonia, Susan and Sylvia (Bingbong). Sandy eventually obtained an M.D. degree from the University of Santo Tomas (UST), Manila and topped the medical board examinations later. Susan became a movie and tv actress of note, who co-starred with the famous Fernando Poe, Jr. in one film. They all moved to the USA in their adulthood. Auntie Juliet married *Arsenio Tito* with whom she had seven children they named Ronaldo (Ronnie), Danilo (Danny), Alma, Esteban (Dudoy), Noel, Rhodora (Baby) and Maria Rosario (Gigit). They all migrated to Australia, except Noel and Rey who migrated to the U.S. and Ronnie who died earlier and Dudoy who remained in the Philippines. Auntie Lucing did not marry. She "worked" in an illicit casino in Pasay City from where she earned a living with which she supported my grandmother. Auntie Nanette worked at the Department of National Defense as a civilian employee. She married *Benjamin Ferraris*, who was a master sergeant at the Philippine Air Force. They had six children, namely; Catherine (Tatine/Cathy), Napoleon (Nonoy/Nap), Maria Victoria (Bubut/Vicki), Josefina (Josie), Maria Helen (Helen), and Maria Cristina (Tina). Uncle Tony did not finish college and married *Concepcion Perez* from Catbalogan, Samar. They had seven children, namely; Antonio Jr. (Bombet), Maria Eileen, Yolanda Jean (Yojean), Josephine Susan (Joy), Procopio Erwin (Joel), Cyr Ian (Ian) and Don Neilsen (Bong). He separated from his family and lived with a common-law wife in Cubao, QC. Auntie Nita completed a nursing course and worked at the Veterans Memorial Hospital on North Avenue, QC for a while. She later immigrated with her son Joel to the USA, where she married and had two more children. Auntie Lulu obtained a B.S. in Pharmacy degree from UST and married *Plaridel Sevilla*, who was a second mate in a local merchant marine ship. They had three children they named Grace (Chucha), Gregory (Binky) and Gerald (Bidim).

Daddy was an intelligent man and could have been a professional with a college degree. His father had enough income to enable sending three of his eight children through college. Daddy studied bachelor of arts in letters at FEU and the UST but did not finish it when he married my mother, who was only 17 years old with Daddy 25 then. He told me that he had several girlfriends before he met my mother.

Mama completed high school only but she was good in mathematics. She was also a good cook. She often helped her mother prepare the meals for her 12 siblings. Nanay, as I fondly called my grandmother, bought and sold jewelry which she got from Bambang, Sta. Cruz, Manila. This was why Mama looked after her siblings when Nanay was out which was often. This situation sometimes caused me and my uncle Raul, who was just four and a half months older than me, to share my mother's breast milk when we were still toddlers. There were other times when we shared Nanay's breast milk, too. In effect, my uncle Raul was my breast brother.

I was the eldest of my parents' four children. My siblings were Eduardo (Eddie), Elizabeth (Beth) and Constancia (Olive). The first saddest moment of my childhood was when Eddie died when I was about two years old. I cried so loudly while Mama held me in her arms as his coffin was being lowered into the grave at the Manila South Cemetery. I wanted to come with my brother in the grave. My second saddest boyhood moment was when I saw my youngest sibling lying down by the wall of the hallway while some of my cousins, Beth and I were playing in the living room of my paternal grandmother's house on G. del Pilar St. She was about two years old when she passed on due to typhoid fever.

My only surviving sibling, Beth, finished only 3rd year in high school. In 1962, at only 16 years old, she worked as a singer at the Aroma Café on Azcarraga St. (now C.M. Recto Ave.) near Avenida Rizal in Sta. Cruz, Manila. One night, she brought home a baby boy, named Bogie (Marion), who was a son of a patient in Tala Leprosarium in Caloocan City and a fellow singer, named *Tomas Alcantara*. She met *Alfredo del Rosario*, who was a guitarist, and married him in 1963 when she was barely 17 years old. They had a daughter (*Lourdes*) and two sons (*Ariel* and *Arnel*), before they separated. She subsequently lived as a common-law wife of *Danilo Paragas* with whom she bore two more children, *Bernardino* and *Teresa*. She fully supported our family during her stint at the Aroma Café since Daddy was in-between jobs then and I was still completing my Radiotelegraph (RTG) Operators Course at the Feati University in Sta. Cruz, Manila. Her support included my daily transportation allowance until I graduated in 1964. Bogie turned out to be our lucky baby because Daddy and I found jobs after we adopted him into our family.

I was in 3rd year high school in 1957 when auntie Lucing agreed to support my schooling by providing my daily allowance. I was not aware then that Daddy may have talked with her into it. I had to live with her at G. del Pilar St., to help in the household. With her were my paternal grandmother, my aunties Lulu and Nita, and my cousins Jun, Sandy, Susan and Cathy. Jun was taking a commerce course at De La Salle University while Sandy was taking medicine at the UST, and Susan and Cathy were enrolled in nursing at St. Paul's College in Manila. All four of my cousins were supported by their parents, including allowance for their meals and laundry. In return for my support from Auntie Lucing, I did my own laundry and volunteered to scrub the floor with a coconut husk. Sometimes, I had tears in my eyes for self-pity while praying in bed before sleeping. My situation at this time and our living condition in Taal St. made me resolve to complete my studies to improve our status. Uncle Naning shared a private joke with me as the two of us being in the Top 10 of my grandmother's "favorite" list! We were at the bottom two. We shared the same interest for singing. He and Jun participated separately in the singing contest of the noontime TV show "Student Canteen", hosted by *Eddie Ilarde, Bobby Ledesma* and *Leila Benitez*. They were both judged Student of the Day. I also participated in the same singing contest later, using *Raul Castillo* as my stage name, but was adjudged only runner-up to the Student of the Day. Raul was my best friend and classmate at the Central Radio and Electronics School (CRES) in Iloilo City in 1960-1961. He taught me how to smoke cigarettes one rainy and cold day when he invited me to have coffee in a carinderia near the CRES and offered a stick of brown Philip Morris cigarette. I felt good after taking a few puffs because of the cocoa flavor of the cigarette. I smoked about 15 sticks a day since then, initially Philip Morris then switching to Newport and Hope cigarettes, which are menthol flavored and cheaper, by the way. I made my first attempt to stop the habit in 1985 which lasted for five years until a co-employee offered me a stick of my first favorite brand, which had a menthol version then. I began smoking again until I finally stopped with a stick left in the pack that I crumpled and threw in the waste basket when I coughed heavily while smoking during a break from the exams given to prospective candidates for the director's position at the office. It was on November 4, 2004. I did not touch a cigarette since then.

My Elementary and High School Years

Mama brought me for enrolment at the Rafael Palma Elementary School (RPES) in San Andres Bukid, Manila in 1949. The only requirement then to be admitted into the first grade was to reach your ear with the opposite hand because many of the birth certificates of children were either burned or not available due to the recent war. Although with an unusually large head (I was sometimes called "Ulo" by my relatives), I was able to satisfy the requirement and began Grade 1 at age six. Our classroom was at the 2nd floor of a house on Amatista St. and our teacher in-charge was Mrs. Bello. She was a kind and fair woman, who once called me and another classmate to the front of the class to hold a book with stretched arms sideways for being talkative at the back of the class. My elementary school years went by without much memories except the time when I asked Mama if I can join the cub scout and told her that I need P2.00 for my uniform. She sadly replied that we do not have the money to afford it. I was so dejected then not knowing the other expenses involved during camping. I merely accepted Mama's decision without question. In Grade 5, Daddy brought me to Bayombong, Nueva Vizcaya and enrolled me at the St. Mary's Academy beginning in June 1953. I now think that he went there to seek greener pastures in employment. Auntie Juliet and her family lived there and they had a motorcycle sales store. Also, Uncle Seniong had a brother who drove trucks for hauling logged woods in Nueva Vizcaya and Isabela. Daddy was hoping he could get a job in either of the two businesses. Apparently unsuccessful in landing a better job in Nueva Vizcaya/Isabela, he brought me in November 1953 to Iloilo City to look for work and to request Auntie Inday to support my schooling. I resumed my Grade 5 schooling at the Rizal Elementary School in Tanza, near my Auntie Inday's home on Hippodromo St. The only memorable incident that occurred to me here was my after-class fist fight with a classmate behind the school building. It ended with him having more red bruises on his face than me. I think Daddy's initial training of my boxing skills paid off. The best result here was our mutual respect. Daddy went back to Manila before Christmas 1953.

My Grade 6 teacher in-charge was Mr. *P. Vinluan*, who was a gentle big man. I remember four of my classmates in this grade. They were *Asuncion Lero*, who made my heart beat faster, especially when she danced in a field demonstration, *Ernesto (Boy) V. Almario*, who was my boyhood neighbor across Dagonoy St. and *Rosalina (Lina) Gan*. Boy and Lina became my classmates again in 4th year high school (Section 1). The other person was *Teresita (Tess) R. Santos*, who was a tall girl with a pair of big round eyes. She also was my schoolmate at the Manuel Roxas High School, renamed Manila High School in 1958, at the Mehan Gardens next to the Manila City Hall on Arroceros St.

R. Palma Elem. School, Grade VI, 1954-55, Mr. P. Vinluan, In-charge. (2nd row) A. Lero at left; R. Gan, 4th from right. (4th row) E. Almario at left; B. Soriano, 3rd from right. (5th row) T. Santos, center

My high school days began when Mama brought me to the M. A. Roxas High School for the start of school in June 1955. I was in tears when she left me at the school. Later in the afternoon after classes, I told Mama that I do not want to go to school anymore. I was overwhelmed by the thought of being far away from home. She told me to tell my father when he arrives from taxi driving. I know Daddy was a strict person and I was afraid that he would castigate me if I told him about my decision. I did not.

My first year in high school was the adjustment period. It was really a difficult time because my classes were held in three different venues. Most of my subjects were at the main campus in the Mehan Gardens. I had some subjects assigned at the National Radio School and Institute of Technology near the Ideal Theater in Sta. Cruz, Manila. My algebra class was held here and I easily liked it. Due to our low economic status, I had only a 72-page lined notebook which I divided into two lengthwise sections and drew lines in-between the lines to increase the writing space. This may be the reason for my small penmanship today. I folded the notebook lengthwise and carried it in my back pocket because I did not have a bag and books to school. My mechanical drawing class was held at the Intramuros Annex, with Mr. *Pedro Picasso* as our teacher.

A new education curriculum in high school, dubbed the 2-2 Plan, began in 1955. Students had the option to choose as an elective, a pre-collegiate subject or vocational subject. I chose the latter and I enrolled in the woodworking class in my first year at the Intramuros campus. It was during this class when I met an accident. Being totally ignorant of carpentry, I tried to polish a short piece of rectangular wood on the electric planer. The wood flew out after touching the blade, which caught and cut half of the tip of my right middle finger. I was brought to the PGH where my finger was treated. My mother was in tears when I went home that day, saying to me to take care of myself next time. My 2nd to 4th year elective subjects were sheet metal working, music and retail store management, respectively. I cannot remember anything about the rest of my elective subjects, except music and attending the other classes where I did not learn much. These elective subjects showed my incompetence in manual skills and financial aptitude, which would later dictate the course of my life.

Mr. Pedro Picasso is at center while I am seated 2nd from the right.

I was in Section 18 in my 3rd year high school, the 3rd best section in the Mehan Gardens campus. Incidentally, Sections 1 to 15 were in the Canonigo (Paco) campus, which retained its name after 1958. Our teacher in-charge was Mrs. *Filomena Hernandez*, who shed motherly tears on the last day of the school year. I had my first real crush in one of my classmates in Sec. 18. Her name was *Maria Francisca A. Acasio*, who always had long threaded hair. I think she had a mutual feeling towards me because I often caught her glancing at me from her front seat two rows to my right. Being young and inexperienced in this matter, I only made "ligaw-tingin". I did not express my feeling or talk with her despite her sending me a Christmas card with the words "Thinking of You"! I was "torpe" (dumb) then. I think she was one of the reasons why Boy Almario did not want to see us 49 years later when I, together with three of my high school batchmates, went to his home in Malabon, Rizal to invite him to our golden graduation anniversary celebration. He was with me and some other classmates when we went to Ma. Francisca's house on Sta. Mesa Boulevard, now E. Rodriguez Avenue Ext., one afternoon after class. Another reason may be because I outdid him in the school standings. Perhaps, he could not accept that a son of a taxi driver could best him in school. His father was a lawyer and they had a car.

Living with my relatives in G. del Pilar St. induced, actually forced, me to study my lessons at night before going to bed because I have no friends around to distract me. I did not have any textbook, so I went to the library frequently to borrow books that I needed for my lessons. This resulted in my high grades (80's to 90's) in mathematics, geometry, physics, and English. My math teacher, Mrs. *Maloles*, once told me that if I remained her top student after the 2nd departmental examinations, she would give me the highest possible grade of 95%. During the 3rd exams, my seatmate (*Renato C. Esquivel*) asked for the answer to one or two questions in math and I readily gave it. He topped the exams and Mrs. Maloles gave him 95% while I got only 90%. This taught me a lesson to be selective in being a generous friend. Mrs. Maloles gave me the highest grade after the 4th exams and nominated me to participate in the qualifying contest to choose the school's participant in an inter-scholastic math competition. I bested two other aspirants in the finals and became the school representative. Unfortunately, I did not win in the competition proper. It was much later, about 45 years after, that I learned the names of my two math competitors when they individually told me on Facebook of their disappointment for their loss to me in that contest. They were *Dionisio Claridad*, a.k.a. *John Claridad*, now a U.S. citizen, who became a lawyer, and *Gabriel M. Abad, Jr.*, also a U.S. citizen now. They are also both retired now.

I liked so much our music teacher in my 3rd year elective subject, who was Mrs. *Rosa Tan*. She was so good at her subject and piano playing. However, she only briefly taught us to read musical notes.

I also think that she was a good mother. I did not see her having a bad mood during the year. Like a true music lover, she had always a pleasant look and disposition. In the middle of the course, she directed us to perform in an operetta of the Merchant of Venice by Shakespeare. She assigned me the role of Shylock with *Fleurdeliz S.J. Abellon* as Portia and *Armando Capistrano* as Bassanio. I was happy to hear the audience laugh when I imitated Portia with her line "I can fly, I can fly" while dancing around like a butterfly fluffing my arms like its wings! Fleurdeliz became my classmate in the 4th year. I heard later that Armando left the country. After the school year ended, I was excited to receive my report card which indicated that I have been promoted to the 4th year, Section 1, the cream section of the school!

I did not expect anything special when school opened in June 1958. I know I will be with the cream of the crop of M.A. Roxas High School, which was renamed as Manila High School (MHS) in October 1958, so I thought I have to study hard to keep pace with my classmates. Our teacher in-charge was Ms. *Araceli Arceo* who was our teacher in Filipino. I did not make it to the Honor Roll after the first departmental examinations, which were held every two and a half months, because I did not pass Ms. Arceo's subject. She gave me a 70% grade. I must admit that I was weakest in Filipino because Daddy used to speak to me in English when I was a boy (This was also my signal to know when he was mad at me - he scolded me in English). Ms. Arceo reminded me to study harder in her subject because she knew I was doing good in the other subjects. I kept in mind her admonition since then. I made it to the top ten in the succeeding Honor Rolls until the fourth or last exams. I was named the 2nd Honorable Mention, behind *Rosalio P. Torres* (Valedictorian), *Consuelito U. Legaspi* (Salutatorian), and *Corazon C. Rodriguez* (1st Honorable Mention). Next to me were *Alfredo P. Larracas, Walter B. Feir, Purita V.* Manlapaz, *Rodolfo R. Obiniana, Benjamin F. Canovas* and *Rolando G. Sobretodo* as 3rd through 8th Honorable Mention, respectively.

The Rafael Palma Elementary School named me its Most Outstanding Alumnus of 1959. I was invited to receive my ribbon for the honor during its graduation ceremonies but I asked auntie Lucing to pin it on me. I later regretted why I did not choose Mama Aba to do it.

MHS Class IV – 1

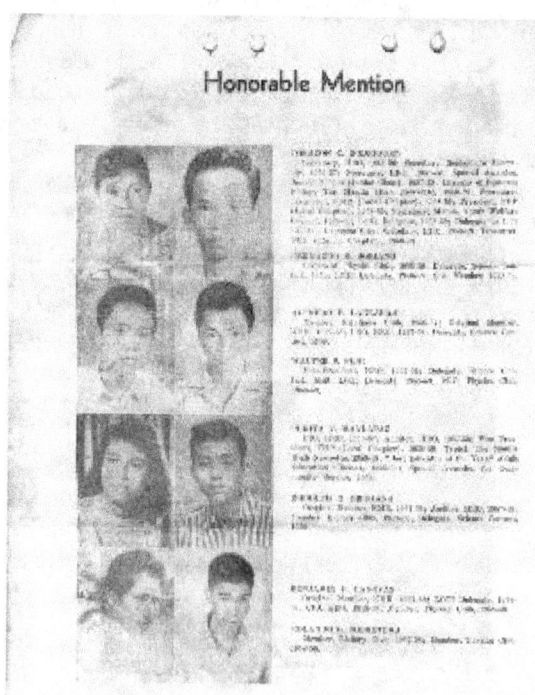

MHS 1959 Honorable Mention

My first experience as an earner was as a newspaper boy when I was in 3rd year high school in 1957-1958. This was when my auntie Lucing, who was unmarried and childless, agreed to support my schooling. She had no regular means of livelihood but she "earned" a living at some casino in Pasay City. Apparently, she had somehow mastered the art of winning in a monte game. I sold newspapers to augment the allowance she gave me. Later, I worked as a waiter at the casino from 7 p.m. thru 7 a.m., with 2 pesos as my daily wage. Sometimes, I get a tip from some winning players and this helped me a lot for my studies. I was just starting my RTO course at Feati U then and went to school from 3:30 p.m. to 5:30 p.m. I also worked briefly as a bootblack at a stand along San Andres St. near Taal St. in Singalong, Manila.

After graduating from high school, I tried to find employment at the Philippine Air Force (PAF) by seeking the assistance of a family friend who knew a colonel there. I went to the PAF headquarters at Nichols Air Base in Pasay City and saw the officer who referred me to the recruitment office, where I took a physical exam and aptitude test. I was told that I did not meet the height requirement, which is 5 ft. 4 in. I stood at 5 ft. 3 in. only. Being unsuccessful in my first attempt to find a job, I thought of applying to the U.S. Navy. I heard that it is accepting applications at its station in Sangley Point, Cavite City. I lost interest in this because I did not have the funds to travel to Cavite City.

My College Studies and Employment

I enrolled in the Radiotelephone Operators (RTO) Course at the Feati University in Sta. Cruz, Manila in June 1959 as a partial scholar, having been awarded the 2nd Honorable Mention at the Manila High School. The scholarship gave me a 25% discount in the tuition fees. I did not study from November 1959 to March 1960 as Daddy brought me to Iloilo City to seek the help of his elder sister for my studies. From June 1960 to March 1961, I completed my RTO Course and began the Radiotelegraph (RTG) Operators Course at the CRES in Iloilo City. I learned how to type fast, use the buzz key and telegraph sounder in sending Morse codes in the CRES. At the end of the school year, I graduated from the RTO course and the CRES bestowed on me the Highest Honors award, which I found later to be the same as a *summa cum laude* in a university. My auntie Inday (Dr. Socorro S. Valdez), who was then the City Health Officer in Jaro, Iloilo, supported me financially throughout my studies in Iloilo City. I complemented her support by participating in various amateur singing contests, such as Halintang sa Kadungganan (Tawag ng Tanghalan) and Darigold Jamboree. I won first place a couple of times when I received 20 pesos but shared 25% with the guitarist later as a gesture of gratitude. It was at Darigold Jamboree where I competed and lost against *Boy Leonardo*, who later on became a champion for several months at "Tawag" in Manila, hosted by Lopito and Patsy.

I had another accident in 1959. When I was playing with my uncle Peter, who visited us from Seattle, Washington, I jumped from the 4-foot post of the fence of our house in Taal St. I did not notice that a vine caught my left foot and broke my balance when I jumped. I landed on the ground hands first and when I looked at my left arm, it was fractured at the lower middle part! I had to endure more than six months with my arm in plaster cast.

My first real job came in 1961 when Daddy called his cousin, who worked as an accountant at a plastics factory in Balintawak, Caloocan City. I was employed at the Solar Plastics Corp. along E. de los Santos Ave., near the "Unang Sigaw" Bonifacio monument, with a 4-peso daily wage. I literally sweated for that amount since I was assigned at the packing department, which was at the 2nd floor of the factory that had no ceiling but the tin roof above us. I do not recall how many times I wrung my face towel of my sweat throughout the day! I worked from 7 a.m. to 4 p.m. with a one-hour lunch break from Monday to Saturday, with payday on the last day.

I was sickly when I was a boy, afflicted with asthma. Daddy and Mama would often bring Beth and me to Luneta, now named Rizal Park, to get sea air, which they said is good for asthmatic people. I had frequent bouts with influenza. Sometimes, when my fever was so high, I would dream that my soul had left my body and was circling above me during sleep. When it finally returned to my body, I would feel alright the following morning and I would ask Mama to prepare a hot calamansi drink and sinigang for lunch.

I went back to school at Feati University to resume my RTG Course in June 1963 while I was still employed at the Solar Plastics Corp. I once caught a heavy downpour while going to school, which resulted to my being sick with coughs and fever the following day. The coughing lasted for three days until I could hardly breathe when I called Mama to bring me to the Medicus Clinic in Pasay City, which was her usual place for our medical treatment. It turned out that I had bronchopneumonia.

I had to endure two daily injections on my upper arms during my confinement for three days until the coughing subsided.

When I was about to complete the course in November 1963 (My photo at right shows the RTG course graduate), our instructor in radio communications, Mr. *Lorenzo Tolentino*, informed us that the Weather Bureau where he works, is giving an entrance examination soon for a training course for meteorological observers and communicators. Some of my classmates invited me to take the entrance exams at the Weather Bureau. I readily agreed to take our chance there and we individually went to the Bureau in January 1964 to

apply for the exam. After several days, I went to check if my name was included in the list of examinees posted on the bulletin board of the Bureau. I observed that many of the applicants were already in their 3rd year in college while I was just a graduate of a 2-year vocational course. There were 109 applicants and only 30 will be selected to undergo the training course that was to start in May. When I went home from the Bureau, I passed by the Quiapo church and prayed to St. Jude. I learned earlier that the saint was the patron of "impossible" wishes. I prayed that I will be the 30th passer in the exam. The exams took place in March. Among the proctors then were *Claro S. Doctor* and *Rolu P. Encarnacion*. I did not know then that they would have significant roles in my career and personal life at the Bureau later. A few days later, some of my RTG classmates congratulated me for passing the exam and, thank God, I was indeed the 30th ... from the bottom of the list of passers!

The Weather Bureau, which was originally the *Observatorio de Meteorologico de Manila* (OMM) or Manila Observatory, was housed in the Marsman Building, just outside Pier 15 in the Port Area, at the back of the Manila Hotel in 1964. It occupied the 3rd and 4th floors of the building. The Administrative and Budget Divisions, together with the Records and Personnel Sections, and the Climatology and Agrometeorology Division were located on the 3rd floor. The Training and Research Division, the Weather Division, the Synoptic Division, the Hydrometeorological Division and the Astronomy Division were all on the 4th floor, together with the printing unit under the last division. The meteorological radar was also on the 4th floor. It was then the only weather surveillance radar in the country. The weather observation station was on the roof deck, where the instruments shelter that contained the dry and wet-bulb thermometers and hygrometer, and the tipping-bucket and standard rain gauge were installed. The Geophysics Division was in a compound with the motor pool in an area owned by the National Waterworks and Sewerage Authority (NAWASA) in Old Balara, Diliman, Quezon City. The Astronomical Observatory with its 29-inch reflector telescope was in the UP Diliman campus near Katipunan Road. There were around 50 meteorological observation stations and a few seismological observations stations scattered around the country.

I learned to love the Weather Bureau because it allowed me to be employed there without any letter of recommendation from a politician, who is called padrino in our local dialect. I went to the Bureau to apply for the entrance exams with only my ball pen. I was received by Mr. *Fernando Tienzo, Sr.*, who was the assistant chief of the Research and Training Division. He gave me the application form to fill in without any question. I learned later when I was already employed that the employees were quite friendly and were like a small family (There were only about 100 personnel then at the Central Office in Manila). In fact, I concluded much later that the office was ran like a small business enterprise, without much standard operating procedures or SOP. This inspired me in my waning years (2002-2008) with the agency to prepare single-handedly, with the assistance of a consultancy agency in Makati City, the Quality Management Systems (QMS), under the International Organization for Standardization (ISO), of the PAGASA. I wrote the six QMS manuals of the agency.

Our classes for the Meteorological Observers-Communicators Training Course (MOCTC) started on May 18, 1964. The theoretical phase of the course ended in September while the practical phase lasted up to October. The course was designed for personnel who will be assigned in the field stations to make weather observations and transmit weather reports in coded messages to the weather forecasting center in Manila. They were later required to have a radio operator's license from the Bureau of Telecommunications. The weather reports were plotted on synoptic charts, together with other weather data from neighboring countries in Asia and the western and southwestern Pacific Ocean, including Australia and New Zealand. Plotted maps were then hand-analyzed by meteorologists who drew isobars (lines of equal atmospheric pressure), isotherms (lines of equal temperature), and isogons (lines of equal wind direction), to delineate high- and low-pressure areas and other atmospheric systems such as fronts, ridges, and troughs. The intertropical convergence zone (ITCZ) is a trough, which is an elongated zone of low pressure characterized by cloudiness and rainfall. The ITCZ meanders in the tropics and is the area where the northeast and southeast trade winds converge. A tropical cyclone is a generic term used by meteorologists to describe a rotating, organized system of clouds and thunderstorms, that originates over tropical or subtropical waters and has closed, low-level wind circulation.

Figure 1: Office building of the Manila Observatory on Padre Faura Street before 10 February 1945.

I ranked first after the theoretical phase and I did not undergo the practical phase because I was extended an Emergency Temporary Worker (ETW) appointment on September 22, 1964 to work as a communicator for an international conference to be hosted by the Weather Bureau in the University of the Philippines Diliman campus in Quezon City. After the conference, Mr. *Hugo de la Cruz*, Chief of the Technical Services Division (TSD), offered to extend me a regular appointment in the Bureau. He asked whether I was willing to be assigned in the field or at the Central Office in Manila. I answered that I wanted to continue my college studies in Manila, so he gave me a Teletype Operator position at the Communications Section under the TSD. My appointment was approved on November 2, 1964 and my starting salary was P150 a month.

L-R: Felino Bartolome, Edgardo Caisip, Teodoro Ambas, Bernardo Soriano, Jr., Oscar Nazareno, Carmelito Calimbas, Joel Contreras, Celso Arellano, and Rufino Lanaca

It was one Sunday in 1964 when my auntie Amanda sent me to bring food for a certain *Gertrudes* (*Truding*), who was a girlfriend of uncle Raul, at the PGH. I learned that she had pancreatic cancer and nobody was taking care of her in the hospital. My errand to bring food to her became more frequent until I felt compassion for her and it became voluntary to me. When she needed blood for her surgeries, I went to the Red Cross on Isaac Peral Ave., now UN Ave., to ask for donation or buy one or two bags of blood. She was allowed to go home in F. Torres St. when she got a little better. However, after a few weeks, her health got worse again and we brought her back to PGH. She expired during one early morning and I went home in tears, telling Mama that Truding had died. It was then when someone knocked at our door and when I opened it, it was a man who had followed me from the hospital offering his funeral services to us. Mama told the man that she will make the funeral arrangements in the morning. My experience watching Truding at night at the PGH taught me how important the role of nurses is. They are the ones implementing the doctors' orders and taking care of the patients while in the hospital. They have to check their emotions in seeing many people suffer or die during their shift.

Mr. de la Cruz assigned me to oversee the three on-the-job trainees (OJTs) at the switchboard, which was operated by my MOCTC classmate, Ms. *Felisa Quional*. One of the OJTs was a lady, who attracted my attention earlier at the Feati University when I was taking the RTG course because she had that teased up hairdo that was in fashion at the time. When I requested for her name, I had to make her repeat it because it was a long one and I did not get it the first time. She turned out to be *Consolacion (Connie/Conching) R. Altavano*, who finished an RTO course from Feati U and was hoping to be employed anywhere. I did not have an initial interest in her because of my timidity with the opposite sex. She was left on the switchboard at the end of Fely's shift at 2 PM and I continued to supervise her from my room adjacent to the switchboard and Mr. de la Cruz's office as I also served as his aide. Connie was a beautiful and sexy lady with smooth, white and shapely legs which were accentuated by her high heel shoes and above-the-knees skirt (Most women did not wear pants then). Her pea-sized mole on her lower left cheek also served as additional attraction to her. I even egged on my co-employee, *Alfonso Obiana*, who was taller, fair-complexioned and more handsome, to try his luck with her but he did not oblige me.

Connie and I became casual friends since then and we left the office together at 5 PM. I learned that she was born in Legazpi City, the 11th among 12 children of *Aurelio Altavano* and *Segundina Rapirap*. We sometimes rode the Matorco, a double-deck bus with an open upper deck that plied along the Roxas Boulevard from Luneta to Baclaran and back. At first, I did not have any romantic intention for her but our constant conversation during and after office hours when I ride with her on the Matorco and to her residence in San Francisco del Monte (SFDM) in Quezon City, drew her closer to me. Connie told me that her father passed away when she was only two years old. This left her mother taking care of her children by being a laundry woman. Due to this, she and her seven sisters learned many house chores, including sewing torn clothes with needle and thread. She also told me that her older sisters *Mamerta (Mer) Altavano* and *Catalina (Kate) Altavano* supported her schooling at the Feati U by providing her tuition fees and daily transportation allowance.

I cannot forget our first date. It was on March 17, 1965 when we arranged a foursome movie date with her co-trainee, *Erlinda Pamintuan*, and *Jose de Guzman*, who was Mr. de la Cruz's nephew and my co-employee. We watched the "Sound of Music", which was playing at the Ideal Theater along Rizal Avenue, near Carriedo St. We all sat in the orchestra section. As I was still inexperienced in matters of the heart and my feeling for her grew more intense, I gathered enough courage to write her a love letter one day. She had no reaction after reading it but she continued her friendly demeanor towards me. I did not know what it meant then.

One day in 1965, Fely received a call from the Philippine Long Distance Telephone (PLDT) Co. which was on strike, asking if she knows of any telephone operator who would be willing to work as a casual employee during the strike. She immediately gave the name of Connie, who was hired soon after at the PLDT. Connie was an intense worker and became a regular employee there after the strike. She rose from the ranks and became a section supervisor after two or three years.

My first appointment in the Weather Bureau was as a Teletype Operator, using the teletype machines and radiophones (Photo at right), at the Communications Section under Mr. Lorenzo Tolentino, from November 2, 1964 to August 31, 1965. I received weather reports and administrative messages from the 56 weather 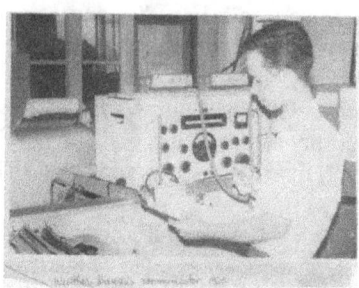 stations in the field. My training in school using the manual typewriter was a valuable asset that enabled me to receive reports and messages with the machine in a single-sending mode. The chief of the Mindanao Regional Office, Mr. *Farnacio Luistro*, once expressed his appreciation of my efficiency in receiving messages. I was also assigned in the graveyard shift (10 PM – 6 AM), since weather observation is a 24/7 work. Although weather reports are supposed to be done and transmitted every three hours, I could not sleep on tables in the office in between observation and transmission times of weather reports. I often just listened to or talked with the observers in the field who were also awake.

Mr. *Juanito F. Lirios*, Chief of the Hydrometeorological Division, recruited me on September 1, 1965, to be one of the staff members of an NSDB-funded project "Rainfall intensity estimation using a radar". The radar in Manila was the only weather surveillance radar in the Philippines at that time. I was given a Science Aide III position with a monthly salary of P230. With me in the project were *Valerio (Val) L. Saniel* and *Rogelio (Roger) A. Wamelda*. Val was our more senior member. We took photographs of cloud echoes on the radar screen when there were tropical cyclones within the Philippine Area of Responsibility (PAR). The photos have various colors with green representing the rain areas. The project ended on August 31, 1967. Immediately after the project, I was hired again as an ETW with a daily wage of Php6.25. I was given this position while waiting for my regular appointment as a Sr. Weather

Observer to be noted by the Civil Service Commission. I remained in the Hydromet. Division.

My relationship with Connie was unclear since she was employed at the PLDT. She has not given me her sweet "Yes". I am not sure now if I ever asked for it! In spite of this, I continued courting her, fetching her from work at the Lawton office when she worked the second shift (2 pm to 10 PM) and riding with her to her home in SFDM. I enrolled in the Bachelor of Science in Electrical Engineering (BSEE) course in Feati U in November 1965 and my classes were from 5:30 PM to 9:30 PM. This meant that I went home past midnight daily during the week. I usually woke up at 6 AM in order not to be late for work at 8 AM.

Our relationship was unstable with frequent breakups and reconciliations, depending on my moods, which was volatile. Sometimes, I wanted her to give me a sign that she loved me, too, but she did not show it. We had our longest break from 1966 to 1967. We did not have any communication at all during this period. Before Christmas of 1967, I went to Bayombong, Nueva Vizcaya to spend the holidays with my relatives there. While there, I missed her and decided to send her a telegram, greeting her "Merry Christmas". When I went back to Manila the following January, I called her at her office to ask if she received the telegram. She acknowledged its receipt and agreed for me to fetch her again that night.

We were "officially" on when she whispered to me with her body against my arm, while on the bus from a visit to the Antipolo church of Our Lady of Good Voyage, that she was mine. That was the happiest day of my young life, having my first real girlfriend!

I must confess that I had another girlfriend, named *Lourdes (Ludy) Pineda*, during our longest break but she was unlike Connie, who was more beautiful and open. Ludy was still quite young and seemed inexperienced, too. Her father was my co-employee at the Weather Bureau, where she tended the canteen. I was disappointed when we had our first and only date to a show at the Araneta Coliseum in Cubao, Quezon City with her kid brother as chaperon. Our relationship did not last long after that.

Luzon Earthquake of 1968 and Our Wedding

I was asleep on August 2, 1968, when a strong shaking woke me up and caused the large (2.5' x 3.5'), heavy mirror hanging aslant above me to fall, fortunately missing me. I learned later that a 7.3 magnitude or intensity VIII earthquake occurred, with its epicenter in Casiguran, Aurora province. A small non-destructive *tsunami* was generated and at least 207 people were killed, mostly from the 6-storey Ruby Tower in Sta. Cruz, Manila that collapsed like an accordion (Photo below). Connie and I went to see the Ruby Tower a few days later.

Ruby Tower in Sta. Cruz, Manila collapsed like an accordion after the intensity VIII earthquake on August 2, 1968.

In September 1968, I proposed to Connie to get married to avoid another breakup. She did not object and we agreed to set up a "pamanhikan". I immediately informed Mama Aba about our plan and told her that we will go to ask for Connie's hand in marriage in November. Being a loving mother, Mama agreed without hesitation. Daddy did not say a word of disapproval and came with us to SFDM. Connie's mother, *Segundina R. Altavano*, was already in their house on Morato St. in SFDM when we arrived in the evening with some dinner foods. She was a gentle woman, who welcomed us into their home, which Connie shared with her older sisters *Mamerta (Mer/Min)* and *Catalina (Kate)* and older cousin *Salvacion (Bing) Ochoa*. Mer and Bing worked at the National Treasurer's Office, where Bing was the National Cashier. Kate worked in a department store in the Quiapo underpass as a saleslady.

Connie and I immediately started preparations for our wedding. I was in the 3rd year of my BSEE course when we decided to get married. I continued to enroll for the second semester of 1968-1969, in spite of our forthcoming wedding. I was disappointed to get only three 3-unit subjects for the semester because of late enrolment and lack of funds. I was reserving my money for the wedding.

Connie and I were married at the St. Anthony parish church in Singalong, Manila on January 11, 1969. Our wedding entourage were composed of the following: *Luz M. Mendoza, Salome Bunao*, Atty. *Rosendo Alcalde* and *Silverio Baranda* as our principal sponsors, and *Danilo S. Tito* and *Antonia Magpantay* as our best man and bridesmaid, respectively. Luz was my aunt, Mrs. Bunao was Connie's officemate, Atty. Alcalde was my aunt Nieveng's husband, Silverio was Connie's uncle, Danilo was my cousin, and Antonia was another officemate of Connie. I cannot remember the names of our flower girls and ring bearer, as we lost the copy of our wedding invitation long ago. I must give due credit to my friend and officemate, *Juanito E. Lucas*, who took all our wedding pictures.

L-R: A. Magpantay, S. Bunao, L. Mendoza, Connie, Bernie, R. Alcalde, S. Baranda, D. Tito (man behind was my MOCTC classmate, E. Caisip, the original photobomber; he was in our many photos but always photobombing)

Wedding at St. Anthony Church, Singalong, Manila

We had our wedding reception at Max's Restaurant in Ermita, Manila, where half of its famous fried chicken cost only Php2.50 then. Due to lack of funds, we did not go on a honeymoon but instead stayed at home on Luciana St. in Santol, Sta. Mesa, Manila, where we had rented an apartment from Connie's officemate, *Gloria Alonzo*. We lived here with Connie's sister, who was still single and 11 years older. Her name is *Remedios Altavano*, fondly called by her siblings as Doy or Edios. We stayed in Santol for a few months until we moved to Chico St. in Project 2, Quezon City. We rented the lower floor of the house owned by Mr. and Mrs. *Joel Cadaing*. Mrs. Cadaing used to work at PLDT, too.

Wedding reception at Max's Ermita, Manila

In March 1969, the Weather Bureau transferred to its new office at 1424 Quezon City Development Bank (QCDB) Building, Quezon Boulevard Ext., Diliman, Quezon City, next to the intersection of South and West Avenues, 9 kilometers northeast of Manila. The building was beside the television antenna tower of IBC Channel 13. The Bureau rented the 3rd floor through the 10th floor of the main building and the entire three floors of its annex. On the 11th floor was the canteen of the building. The Office of the Director was on the 3rd floor, shared with the bank's administration staff. The 4th floor was occupied by the Administrative Division with its Personnel and Records Sections while the cashier's office and the Procurement Section were on the 5th floor. The Budget and Accounting Divisions were on the 6th floor and the

Hydrometeorological Division was assigned on the 7th floor together with the UNDP experts' office. The Climatology and Agrometeorology Division and the Luzon Regional Office were on 8th floor while the Institute of Meteorology (IM) was on the 9th floor. The (Weather) Forecasting Division was on the 10th floor with the Flood Forecasting Center under the Hydromet. Division. The Field Operations Center was on the 3rd floor of the annex while the Engineering and Maintenance Division was on its 2nd floor. The entire ground floor of the annex building was occupied by two training rooms and the Training Section under the IM. The Bureau canteen, which was operated by the Weather Bureau Employees Consumers Cooperative, Inc. (WBECCI), was beside the stairway from the 3rd floor of the main building. The quadrangle between the two structures also served as parking space.

July 20, 1969 was the most significant date of my and Connie's married life. It was the day when the Apollo 11 astronauts, Neil Armstrong and Michael Collins, made history by being the first humans to set foot on the moon. It was also the date when I was successful in "landing" so that Connie conceived our first child! While Armstrong uttered his famous line "That's one small step for man, one giant leap for mankind", I said to myself "Thank you God, you made me a complete man".

Our first child, a female, was born on April 29, 1970 by caesarian section (CS) at the University of the East Ramon Magsaysay Memorial Center (UERMMC) on E. Rodriguez Avenue Ext., Quezon City. Dr. *Gloria C. Habalo* was the OB-Gynecologist who performed the CS. We named our daughter *Maria Michele* (*Mich*), most probably influenced by the Beatles song with that title and because we thought that it is a sophisticated French name (She excelled later in her college subject in French). Michele or Mich was a beautiful girl with curly hair when she was still a toddler. She looked like a blend of Drew Barrymore and Shirley Temple, the American child actress with the curly hairs. She had a fair complexion like her mother and we adored her so much.

Mich's birth taught me about a mother's sacrifices after bearing her child for nine months. Connie would wake up in the middle of the night to change her diaper, which was then made of plain loin cloth (Pampers or disposable diapers have not been invented yet), or when she was hungry. Connie was still wearing a girdle brace to hold the stitches of her CS wound in place and she had much difficulty in moving during the first months. Knowing this, I sometimes woke up at night to prepare Mich's milk formula and volunteered to wash her dirty diapers in the day during weekends.

There was one time when I came home from the office and Mich, who was then one year old, was running a fever due to a severe cold. Connie was in her office and will be off at 10 PM. Doy, who is Michele's baptismal godmother, brought her to me and I saw her limp and was barely breathing. I was alarmed and initially did not know what to do. Something (her guardian angel?) made me decide to suck her nose to remove the mucous that seemed to block her breathing. After two tries, I was able to extract the mucous from her nostrils and she began to breathe freely again! I simply spitted out the mucous into the lavatory afterwards.

There was another time when we had to bring Mich to UERMM at night due to her very high fever. I was so worried to see that the staff at the emergency room placed her naked on a tub of ice tubes and covered her body with the ice. I did not realize then that it was to lower her body temperature as quickly as possible. Her fever subsided immediately after that and some medicines. We were allowed to go home on the same night.

In the latter part of 1970, when I was in my last units of the BSEE, I applied for the forthcoming Meteorologist Training Course at the Weather Bureau. The course required graduates of collegiate courses in natural or engineering sciences with credits in mathematics up to differential equations and college physics. The course was scheduled to start in February 1971 and I would satisfy the requirements only in March 1971. My application was initially turned down but I wrote a letter to the Director, Dr. *Roman L. Kintanar*, appealing for reconsideration of my application based on that fact and the understanding that I will not be extended an appointment as a meteorologist unless I obtained my college degree. My application was eventually approved and I joined the MTC on February 15, 1971. I completed all my academic requirements for a BSEE degree in October 1971 but received my diploma in April 1972 (Photo at right).

During the MTC, Connie got pregnant with our second child, a male. Dr. Habalo told us that it will be another CS delivery, because of Connie's narrow pelvic bone opening. The delivery was due in February 1972. We selected the 14th, mainly because it is a payday in the office. Being Valentine's Day was another major factor and a cause for celebration when we need funds. For these reasons, we named our son *Christian Bernard*, in reference to the famous surgeon, Dr. *Christiaan Barnard*, who performed the first human heart transplant in South Africa on December 3, 1967, and my first name. My co-trainee, *Reynaldo P. Munda*, volunteered to be his baptismal godfather because we became quite close to each other personally and academically. He and I ranked 1st and 2nd, respectively, at the end of the training course on February 12, 1972.

Christian, who we later nicknamed Chris, was a cute, chubby baby who broadly smiled when I looked at him on his crib or on the straw mat or "banig". He had a large appetite, especially for soup, when he was about one year old. Sometimes, I get tired feeding him. I would ask him if he wanted some more after finishing two plates of rice and soup. He would nod approvingly and said "baw", referring to "sabaw" or soup. He was very sickly when we moved in 1973 to Lanzones St., a few houses from our former Chico St. residence. He often had fever in the 40s, which Dr. Habalo diagnosed as due to primary complex. We thought that it was caused by the water paint powder of the walls of our new residence. We had the walls repainted with enamel paint, but his high fevers continued intermittently. We even had the house blessed by a priest to no avail. These went on until he reached seven, when the fevers stopped.

Our experiences with our two children's growing up health issues made me realize that God is, indeed, good all the time. I was wondering then why He allowed our children to get sick every time Connie or I got a financial boost from the office. I realized that He prepared the financial support for us to cope with the crises that were coming.

I was promoted to a Meteorologist position on January 20, 1972. I was assigned then at the Institute of Meteorology, where Mr. *Catalino P. Arafiles* was the Officer-in-Charge. He was the Chief of the Synoptic Division prior to this. Mr. Arafiles, who we fondly called Tatang, was (and still was up to his retirement in 1993) the only summa cum laude and recipient of the University President's Gold Medal when he graduated with a BSEE from Feati University. My first impression of him was a stern and strict supervisor. He required each employee of the Institute to submit a daily accomplishment report in a form. This made us really work throughout the days of the week. This requirement lasted for many months until it became our second nature. He eventually discontinued it without giving us any reason. He was also a benevolent person, who shared the incentive benefits of the projects he managed to many employees in the agency.

The Institute of Meteorology was part of the 5-year " World Meteorological Organization (WMO) Training and Research Project, Manila" in 1969. With the implementation of the project, the acquisition of an IBM-1130 was realized and computerization in the Bureau came of age (https://en. wikipedia.org/ wiki/PAGASA). The central processing unit, console, input machine and printer of the IBM-1130 occupied an entire room of 4 m x 6 m at the Penthouse of the QCDB building. The eight keypunch machines were located in an adjacent 12 m x 6 m room.

IBM-1130 with peripherals including paper tape reader punch (top) and its console (bottom)

Martial Law in the Philippines

A dark chapter in Philippine history occurred on September 21, 1972, when Presidential Proclamation 1081 was issued by President *Ferdinand E. Marcos*. He appeared on television on September 23 to announce the proclamation suspending the civil rights and imposing military authority in the country. He claimed that it was the last defense against the rising disorder caused by increasingly violent student demonstrations, the alleged threats of communist insurgency by the new Communist Party of the Philippines (CPP), and the Muslim separatist movement of the Moro National Liberation Front (MNLF).

Prior to this, there were spates of bombings in several places in the country. The most notable one was the Plaza Miranda bombing incident on August 21, 1971, during a political campaign rally of the Liberal Party in Quiapo, Manila. There were nine fatalities and 95 injured persons, including prominent politicians like *Gerry Roxas, Eddie Ilarde, Ramon Bagatsing,* and *Jovito Salonga,* who lost an eye and several fingers in the bombing. I had a first-hand experience related to this when tear gas bombs were hurled at student demonstrators near the main campus of the Feati University where I was enrolled in the night classes for my BSEE course. The smokes reached our classroom on the ground floor and greatly disturbed us.

Many people felt pressure during this time because a midnight curfew was imposed on the public. My officemates and I usually had our Friday night winding down activities by going bowling while having a few bottles of beer. Mr. Arafiles and Mr. Lirios were usually with me and *Amadeo (Mading) Balotro, Claro S. Doctor,* and with *Nestor Bautista* as our driver. When it was near midnight, we had to hurry up and leave to avoid getting caught on the road since there were police or military checkpoints in vital areas in the Greater Manila Area.

It was at the height of this era when President Marcos abolished the Weather Bureau and established the Philippine Atmospheric, Geophysical and Astronomical Services Administration or PAGASA, under the authority of the Department of National Defense (DND). Presidential Decree No. 78, s. 1972, enabled the establishment of the new agency. The Secretary of the DND then was *Juan Ponce Enrile* while Gen. *Fidel V. Ramos* headed the Philippine Constabulary. These two would be the principal players in the People Power Revolution on the Epifanio de los Santos Avenue (EDSA) in 1986.

PLAZA MIRANDA BOMBING

Four organizational units initially comprised PAGASA. The National Weather Office (NWO) prepared and promptly issued forecasts and warnings of weather and flood conditions. The National Atmospheric, Geophysical and Astronomical Data Office (NAGADO) undertook the acquisition, collection, quality control, processing, and archiving of atmospheric and allied data. The National Geophysical and Astronomical Office (NGAO) was in charge of observations and studies of seismological and astronomical phenomena, as well as provision of the official time for the country. The National Institute of Atmospheric, Geophysical and Astronomical Sciences (NIAGAS) was responsible for the training of scientists and technical personnel with respect to atmospheric, geophysical and astronomical sciences. In 1977, the Typhoon Moderation Research and Development Office (TMRDO) and the National Flood Forecasting Office (NFFO) were placed under administrative supervision of PAGASA, pursuant to Presidential Decree No. 1149, s. 1977. These units were supported by the Administrative Division (AD), the Finance and Management Division (FMD), and the Engineering and Technical Services Division (ETSD).

Sometime in 1972, Mr. *Cipriano* (*Cip*) *C. Ferraris*, who was our instructor in Hydrometeorology at the MTC, recruited me to be the counterpart meteorologist to the United Nations Development Programme (UNDP) expert Mr. *Milorad Obradovich*, who was originally from Yugoslavia. His field of expertise was in hydrometeorology and my initial assignment was doing the rainfall probability analysis of the Philippines with Mr. Ferraris as my immediate supervisor. Cip and I co-authored my first technical paper, entitled "Maximum rainfall values over

Luzon for durations of 1 and 2 days and return periods from 2 to 50 years", completed in 1973 and published as the Weather Bureau Technical Series No. 19, 25 pp.

Mr. Obradovich nominated me and *Cesar Bernardino*, who was a meteorologist-researcher at the Instruments Research and Development Section under the NIAGAS in the Science Garden (SG) on Agham Road in Quezon City, to undergo training on the maintenance of telemetering rain gauges at the Meisei Electric Co. plant in Moriya, Ibaraki Prefecture, Japan from 2 to 24 April 1973. This was my first trip abroad and, like many Filipinos who come to see their relatives off to international travel by the jeeploads, Connie and our two children then, my parents, Connie's two sisters, and our house help, came to see me off at the Manila International Airport in Pasay City. I wore a coat and tie, believing that it was the normal attire when men go abroad.

Mr. Obradovich planned to involve me in a project to install a network of telemetering rain gauges in the Marikina River basin, for flood monitoring and prediction in the tributaries of the Marikina and Pasig Rivers.

We first surveyed possible sites near the Wawa Dam in Montalban and Boso-Boso, Rizal. These places were the upstream areas of the Marikina River. I was assigned to take charge of constructing the cement bases for the rain gauges and housing for its telemetering devices. *Bernardo Dela*, Mading Balotro, and *Marcelino (Mar) Bala* were my teammates for the project. Mading was about 20 years older than me, who was a former trainer of boxers in Sampaloc, Manila. He had big, round eyes that grew much larger when he was angry or excited. For these reasons, he would strike a fighting stance whenever we kidded him that he was better off than we were because he had "malaking kita, daig lang ng duling na doble ang kita" (Had big earnings, second to cross-eyed people who earn double)! He was still an undergraduate when he was in the team. Dela was an experienced construction foreman. He sometimes accepted small construction as his sideline. We called him "putol" due to his surname, which we said is hanging as in de la Cruz or de la Vega. Mar was a mechanical engineering graduate and a meteorologist. He was very helpful to me and we became close friends in the office. He had three sons and made me the baptismal godfather of his youngest son, Michael (Macky/Kekel).

Wawa was a small community, hidden by the Montalban mountains, north of Manila and Quezon City. The site Obradovich chose in the area was behind a 100-meter high hill, downstream of the Wawa Dam. It was on the wall of the mountain which is about 30 meters high, a few meters from the bank of the Montalban-Marikina River. In order to go to the site, we had to climb a trail wide enough for only one person to walk. Anyone who missed a foothold and fell along the trail would most probably suffer serious injuries, if not die. We constructed a 4m x 4m x 3m structure on top of the hill to house the repeater device that would transmit the telemetered rainfall data to the Flood Forecasting Center at the QCDB building. We nearly met a fatal accident during the construction. In one descent using a Willys jeep, which was part of the reparation goods from the U.S. government (we normally used a Toyota Land Cruiser, donated by the UNDP), from the steep dirt road of the hill which had no restraining barricades, our young driver, *Benjamin Tado*, lost control of the vehicle that surged downward rapidly. It was good that he recovered his nerves in time to veer the jeep to the side of the hill. Mading, then a Meteorological Aide, suffered a minor cut on his head which was treated in a nearby hospital in Montalban. I was sitting beside the driver, so I was able to prepare myself by bracing onto the front dashboard and all I suffered were frayed nerves. Ben later told us that he thought he had lost the brakes, but we agreed that it was a case of nervousness or driving inexperience.

We had to pass through Antipolo City and Teresa, Rizal in going to Boso-Boso. There was no asphalt or paved road after Teresa in 1973, and we sometimes rode our Land Cruiser along rivulets with many large stones in San Andres, Rizal. I will never forget one day in October or November when we went during the construction of the bases of the telemetering rain gauge. I was with Mading Balotro and Bernie Dela with *Roger Cuenca*, who was our more experienced driver of the Willys jeep. While we were in Boso-Boso, it rained hard during the afternoon. When we were going home late in the day as the sun set early during this periood, we encountered flooded areas along the way, which caused the jeep to stop. Mading, Dela and I had to push the jeep in thigh-high water and mud, and were happy doing it under the clear, starry sky, until we reached a higher spot for Roger to dry the distributor and start the jeep engine running again.

After we completed installing the two telemetering rain gauge stations in Wawa, Montalban and Boso-Boso, Rizal, we next surveyed a possible station site in the Mountain Province. We searched for a site in the area along the Halsema Highway, the highest highway in the Philippines which, at its highest point in the municipality of Atok, is 2,300 m (7,400 ft) above sea level. We likewise searched in Banaue, Ifugao

Province, among the world-famous rice terraces, for a suitable site. The rice terraces were and still are considered by many Filipinos as the 8th Wonder of the World. I am one of those Filipinos. Lastly, we looked for a possible site near the Salinas Salt Spring in Nueva Vizcaya.

Mr. Obradovich's tenure as a UNDP expert ended in 1976 after he helped lay down the framework for the nationwide flood forecasting system. I returned to the Institute of Meteorology, which has been renamed National Institute of Atmospheric Sciences (NIAS) and later to the National Institute of Atmospheric, Geophysical and Astronomical Sciences or NIAGAS.

I was promoted to Sr. Meteorologist on July 1, 1973 and to Supervising Meteorologist on June 28, 1974. I became Chief Meteorologist on my 9th wedding anniversary on January 11, 1978. I held the Training Division briefly and was promoted again to Weather Specialist on January 18, 1979. These movements were all in NIAS/NIAGAS.

Our third and last child was another male, who was born by CS on November 22, 1973. We named him *Johann Jude*, after the Austrian composer celebrated for his waltzes and operettas, and our patron saint Jude. He was born after we moved to our 4th home on K-6th Street near corner K-J Street, Kamias District, Quezon City. He is our most resilient child, who never had any serious illness during his childhood until his adulthood. The only time I remember that he was sick was when his brother had his fever episode and Jude ran a slight fever, too. One day when I got home from office, Doy who was tending to Chris, removed the ice bag from Chris' head to allow him to take his medicine for his high fever. After a few seconds, Chris began convulsing and the white of his eyes showed. Again, something (Chris' guardian angel?) told me to pour the cold water from the ice bag on his head! Chris immediately stopped convulsing and I rushed him to a nearby hospital on Kamias Road. Doy, who our children later referred to as Nang (short for Ninang), as Mich called her, was a great blessing for us because she allowed Connie and I to work while she took care of our children. She cooked our food, washed our clothes and kept the house clean all the time. In other words, she was our children's second mother.

Connie was already a section supervisor in 1972 at the Del Monte Service Center (DMSC) of the PLDT, beside the Muñoz Market at the corner of Roosevelt Avenue and EDSA, when she and *Filipina (Philip) Santos Olona,* a close friend and co-employee, decided to look for a house and lot to be our permanent home. I accompanied them to Marikina to respond to an ad but they found that the property was not within their financial capability. Someone told them that there are many subdivisions being developed in Novaliches and we went to San Bartolome to look at the one being developed by Mr. *Antonio (Tony) C. Francisco.* We found that his sales terms were acceptable to us. For the 60-sqm house and 209-sqm lot, we have identified and surveyed, the cost was Php35,000 with a Php5,000 down payment, half of which was payable upon application for a housing loan with the Social Security Systems (SSS). At this time, Php5,000 was a large amount that we hardly had. Connie and I secured loans from our office credit unions and equally shared the amount. After paying half of the down payment, Tony transferred the land registration title to us which was our collateral for the loan. The SSS required that half of the house has been built before it approved our loan. Tony took care of the house construction and when it was half-complete, the SSS approved our loan with an amount of Php29,300. When the house was completed in 1973, we paid the other half of the down payment before transferring to it. Tony did not charge us the extra 700 as his friendly gesture to us.

Jude was barely one year old when we moved to our permanent home at the ACF Homes in San Bartolome, Novaliches, Quezon City on December 7, 1974. He is our most cheerful and artistic offspring. I really think now that he got his positive outlook in life from me. He has also been the person who brought Daddy and his aunt *Lolita Altavano*, elder sister of Connie, to the hospital when they expired. He is now taking care of Nang in our house, who suffered a stroke in 2014, rendering the right side of her body paralyzed, and has been bed-ridden since then.

She looks fine now at 86 years old, other than being sometimes irritable due to her condition. Photo at right shows us with Nang (standing behind) at our new home in 1975.

Early in our marriage, Connie and I called each other "Giliw" and "Irog", respectively, as our endearing name for each other. I decided to use these unique terms of endearment for us, instead of the common "papa/daddy" and "mama/mommy", etc. These became "Giwil" and "Irog" when our sons could not pronounce the first term correctly. We call each other these names up to now but our grandchildren have called us "Mamanie" and "Dadad".

I took a rest for three years before I decided to go back to school in 1975 to pursue a Master of Science degree in Meteorology at the University of the Philippines (UP) in Diliman, Quezon City, after graduating in 1972 from BSEE. I thought then that I have to have a graduate degree to rise in the office. I applied for a partial scholarship grant at the PAGASA, which involved at least 20 hours per week of office work. I went to the office in the morning and to the graduate school at the Department of Meteorology (DMO), UP Diliman in the afternoon. At the same time, I was then employed as a part-time instructor at Feati U, teaching meteorology, college algebra and college physics to B.S. in Maritime Transportation students. My classes at UP were held only three days a week, so I spent the days off at the office. My first class in Feati U began at 5:30 PM, so that I had to scamper to take the jeepney rides, sometimes hanging at the running board, as soon as I got off from the office to be on time for my first period at the campus in Echague St. (now Carlos Palanca St.), at the foot of the Ayala Bridge. I had to walk about 300 meters from the jeepney stop under the Quezon Bridge. The vehicular traffic in Manila was not heavy then.

The above situations briefly described my five different roles as a man since I got married and had children. I was a family man and office worker in the morning, a student in the afternoon, a teacher in the early evening, and a husband at night. I also served as an instructor in the in-service training courses of the PAGASA, from 1972 to 1982. I taught hydrometeorology in both the Meteorological Observers Training Course (MOTC) and the MTC, which are the sub-professional and professional levels, respectively, in the meteorological education curriculum of the WMO.

From June 1977 to August 1978, I brought Mich in the morning to the St. Mary's College (SMC) along Panay Avenue, near my office on Quezon Blvd. Ext., where we enrolled her. We took a bus from Holy Cross, referred to by people to denote the Holy Cross Memorial Park in San Bartolome, to West Avenue and jeepney to SMC. I fetched her at noon to bring her back home. I went back in the afternoon to UP when we had classes. We had our first car, a Ford Escort, in August 1978, which we bought from our neighbor in ACF homes. It had the distinctive color of mustard yellow. By then, our two other children were of school age and I used to bring and fetch them and the children of our close friends, neighbors and kumare/kumpare, Mr. and Mrs. *Fernando Z. Olona*, to and from SMC.

Connie decided to go back to school in Feati U in 1975 to take a Bachelor of Science in Business Administration (BSBA) course in the evening. She had her classes at the main campus beside the MacArthur Bridge. As I was then teaching there, we went home together, usually arriving home between 10 and 11 PM. She graduated in 1979 when she was already the supervisor of the testing section of the PLDT.

We were already in Novaliches when a long strike occurred at the PLDT Del Monte Center, beside the Muñoz Market at the corner of EDSA and Roosevelt Avenue in Quezon City. Being a supervisor, Connie was not a member of the employees' union and was not allowed by the strikers to leave the office. She and her staff, who were also non-members of the striking union, were provided by the company with cots, pillows and blankets, drinking water and foods. A helicopter dropped the supplies when it ran out. She would sometimes scale the concrete fence, stepping on empty gasoline barrels in the compound at night with the help of some friends, to see us especially her children, at home. She did this at least three times during the strike, which lasted for over a month.

My second trip abroad was on a Thursday on January 2, 1975 when I was nominated for a fellowship under the WMO to study hydrology at the Colorado State University (CSU) in Fort Collins, Colorado, U.S.A. My plane landed on the same date at the Dulles International Airport, from where I took a limousine bus to report at the National Weather Service (NWS) central office in Washington DC. It was so cold outside and I only had a second-hand overcoat, bought from Bambang, Manila, to protect me besides my suit. I went to a shop near the NWS office to get a pair of gloves, which I found later to be made in the Philippines. After receiving my first stipend and cashing the check a day later, I received my plane ticket to Denver, Colorado and bus ticket to Fort Collins. I arrived in Fort Collins at night when the wind and snow were blowing hard. I was met at the bus depot by a man, who brought me to the house of a couple, who would be my foster guardians until I got settled there. They were a young (30ish) Christian couple who did not have a child at the time. I slept in their basement and stayed in the house for two more nights. On the third day, they drove me to the CSU for enrolment and the College Inn, where I would stay during the quarter.

Besides hydrology, I also took Fortran basic programming as an additional subject at the FSU. Adjusting to a new environment, people and culture was quite difficult at first but I eventually adapted to all of these. I had my first experience of playing in snow like a child outside the College Inn. My classes were in two different buildings which are far from each other and I had to walk on snow without an appropriate footwear. The students were friendly but they were busy minding their own business and were serious most of the time. When I meet people walking, they usually greeted me with the words "Hi, how'r you doing?" nonchalantly without stopping or looking back. It was like, in our local language, "lumalabas sa ilong" statement, that is, they do it as a force of habit and not really mean it.

I met *Jose Campos* of El Salvador and *Julio Vivero* of Panama, who were also on WMO fellowships, during the CSU orientation day before classes began. We shared a room at the College Inn, outside the campus, but Jose moved to another room later due to a minor disagreement with Julio. Jose was a boisterous and happy person while Julio was the quiet and serious type. I did not have any problem with them because I am in-between their characters and, therefore, knew how to adapt with them. Julio gave me two paracetamol pills when I had a fever due to my being soaked in sleet when I biked to a grocery store to buy some personal supplies including toothpaste. I missed my classes for two days and stayed in bed at the Inn. At end of the quarter, which ended in March 1975, I got an A in Fortran and C in hydrology. I parted with Jose and Julio without much ado. We merely said goodbye to each other at the Inn and left Ft. Collins on different dates.

From Ft. Collins, I proceeded to the River Forecast Center (RFC) in Portland, Oregon, where I was trained in river monitoring and forecasting procedures. Mr. *Vail Schermerhon* was my training supervisor. He was assisted by Mr. *David Baumann*, who was my constant companion during the training period. I reported to the center at 9 AM and was allowed to go home at 3 PM. Vail invited me to his home one day and introduced his wife to me. We had dinner there. On my first weekend in Portland, I called my auntie Mimi who lived in Seattle, Washington. I told her I was in Portland and will there for the next three weeks. She invited me to come to Seattle and spend some weekends with her and our relatives there.

On my second weekend at the RFC, auntie Mimi came to fetch me to spend the weekend with her and her daughter. She drove for three hours in coming to Portland. Auntie Mimi was a kind, usually cheerful, and loving lady. She sponsored most of her siblings to come to the U.S., including uncles Butog, Berting, Raul and Nonoy/Inteng and aunties Luz, Chu-chu and Connie. Uncle Celso applied for immigration and, because he was a licensed architect, was readily approved together with his family. We had a happy reunion when auntie Mimi brought me to uncle Celso's residence. I learned that he bought houses, renovated and sold them later. Auntie Mimi fetched me again on my third week in Portland. We had a picnic in a park where seagulls waited for food while hovering in air. They would snatch the food in midair as soon as you threw it to them. She brought us to uncle Butog's house for lunch once. Auntie Mimi took a lot of photos of our activities while we were with her, often directing us how to pose, and gave me at least five one-photo-per-back-to-back page albums later. She was an impulsive shopper, buying many things that are on sale, usually used clothes and package-damaged foods. She had three large sacks full of used clothes in her living and bedroom. She gave me two grocery shopping bags of the clothes, which I stuffed in my two luggage when I went back to the Philippines at the end of my fellowship on April 26, 1975.

I completed my academic requirement for my M.Sc. degree in 1978. Mr. Arafiles re-assigned me then to the National Radiation Center (NRC) in the Science Garden to allow me to concentrate on my studies and finish my master's thesis. The NRC maintains the few solar radiation observation stations in the country. It collects sunshine duration and other components of solar radiation. It sends the data to the U.S. for consolidation with the data of other solar radiation centers worldwide. Together with me at the NRC were *Jesusito (Jess) N. Yunzal, Lorna V. Imperial, Victoria (Vicky) Lamen,* and *Arman Griarte.* Jess was an intellectual, who was also good in mechanical and manual work. We had high level discussions on various issues and topics. He is one of the two persons, the other being *Eliseo (Ely) Salazar,* who have a good potential for advancement in the agency because of their high intellect. Jess supported his only daughter who graduated cum laude with a BS Economics degree at UP Diliman. She later pursued graduate studies at City University of New York (CUNY) and obtained a PhD degree in Economics. Jess' wife, Paz, was then working as a teacher for visually impaired children in Saipan. Ely was an observer, too, at the SG weather station. However, Tatang had not noticed them or he knew of negative feedbacks about them. He often voiced out about Jess working on his motorbike in front of the NRC and Ely and his family occupying the adjacent room of the weather observation station. Jess eventually made me the principal author of the AGSSB Technical Note, titled "Solar Radiation Map of the Philippines, 1999 Update", with him as the co-author. Arman inputted the solar radiation data in a mapping software (later called applications or apps) that automatically drew the solar radiation isolines or *isohels.*

I was still far from completing my thesis due to some distractions in the Garden, after having spent more than a year at the NRC. One distraction is the lack of a supervisor like Mr. Arafiles that allowed us to be lax in our work. Another is, we often exceeded our lunch and coffee breaks, spending them at *Marina Homol*'s carinderia beside the weather observation station. Marina was the spouse of *Alberto Homol,* one of the weather observers at the station.

Mr. Arafiles recalled me to the Central Office in 1979, after one of his Assistant Weather Services Chiefs (AWSCs), Mr. *Catalino (Taleng) P. Alcances,* left the country to accept a United Nations post in Lagos, Nigeria. Taleng died there later for a mysterious reason. He was replaced by Mr. *Bayani S. Lomotan* as the AWSC for research.

This last development led to my finishing my master's thesis quickly. I wrote my thesis using manual typewriters at the office. My thesis adviser was Dr. *Jorge G. de las Alas,* who was the Chairman of the DMO, UP Diliman. Dr. *Rodolfo (Rody) A. de Guzman* was my thesis reader and professor in Physical Meteorology during my academic studies. He reviewed my drafts and greatly improved my thesis. My mother passed away on January 3, 1981 while I was writing my thesis. I had to stop writing my thesis and attend to her funeral services for one week, including the mourning period. In retrospect, I felt guilty for ignoring or turning a deaf ear to Mama's plaint earlier about her feeling a tightness on her breast. I should have brought her to a heart specialist for medication of her high blood pressure that caused her to suffer a stroke that led to her demise. It was the saddest moment of my life! I just arrived at the office from home that day when my officemate informed me that my father called. A few minutes later, the phone rang and my father said that I have to come to the PGH immediately. I arrived there when Mama was lying unconscious in the emergency room. I was informed that she had already moved her vowels, which is a sign that she was already gone. I still hoped and prayed that she would recover but she was pronounced dead after a few minutes. I cannot remember now how I reacted about the pronouncement of my mother's death. Without justifying my negligence then, I could not afford extra expenses because my salary was enough only for my family and our car (Having a car was like having another family). This was the same reason that may have caused Daddy's enlargement of the heart to worsen when he moved in our house after Mama's death. I brought him only to the emergency room of the Veterans Memorial Hospital (VMH) when he felt weak. I thought they only did some palliative remedies there. His heart condition did not improve and I had to bring him back to the VMH several times until he stopped complaining about it. On November 4, 1997, our youngest child, Jude, called me at the office to tell me that he brought daddy to the Bernardino Hospital. I hurriedly drove to the hospital to find the limp body of my father in the ER. I tried to revive him by pounding on his chest several times and saying "Dad, please wake up!" to no avail. My hero and inspiration had passed away.

When I was ready to finalize my thesis, I went to the DMO to type it on an electric typewriter. I was familiar with the machine because I was a teletype operator when I first worked in the Weather Bureau. However, I had to do it on stencils, which was tedious, especially when I made typing errors and had to use a correction fluid to undo the errors. I was very happy after the defense of my thesis, without much revision. Dr. de Guzman consequently left for the WMO headquarters in Geneva, Switzerland with his family to begin his new career there.

NIAGAS had two AWSCs prior to 1979, one for research and the other for training. The AWSC for training was Atty. *Wilfredo V. Garcia*, who succumbed to a heart attack while giving a lecture on administrative procedures to the MOTC class at the Science Garden. He was rushed to the nearby East Avenue Medical Center but was dead on arrival.

In 1979, Tatang called Claro, Rolu and me to sit in the area around his office on the penthouse of the QCDB building. The office of Atty. Garcia was next to his office, which he kept locked after the death of Atty. Garcia. Opposite this room, across the building on the Quezon Blvd. side was Mr. Lomotan's room. I stayed in the adjoining room, where the staff of the AWSCs were seated. Claro was then the counterpart meteorologist at the UNDP experts' office on the 8th floor while Rolu was at the Geophysics and Atmospheric Research Division (GARD) in the Science Garden. Claro opted to stay at the 8th floor while Rolu transferred to Mr. Lomotan's room.

Tatang forwarded correspondences to us individually and asked us to draft their reply. There was still no personal computer then and we used the yellow ruled pad in drafting our replies. It was at this time when *Rosana (Toots) de la Cruz* came into my life. She efficiently typed my drafts. Toots was a beautiful, fair-skinned, young lady who possessed a unique, high-pitched voice, like the character Matutina in the popular tv sitcom John en Marsha. One would not mistake her when she laughs loudly. She was very good in office-keeping as she keeps everything in the office neatly in their proper place. It is no wonder because she completed a medical secretarial course before she worked with PAGASA. She would eventually become my secretary and Connie's wedding goddaughter, until I retired.

We did not know then that Tatang was trying to find out how well we could prepare the replies. I observed that he made many corrections on my first draft, almost completely covering the whole page with his red ink writing, which was quite difficult to read at first as he had a penmanship like that of many doctors of medicine. Once, he called me to his office to point out my "mistakes" and advised me to read previous correspondences prepared by Atty. Garcia and to note his style. Then, slowly the red marks began to decrease and completely disappeared eventually. When he apparently had determined who among us three was most prepared, he called me and Mr. *Gaudioso R. Tabamo*, who was more senior that I, to his office one day. He told us that he was considering to recommend his AWSC (for training) and he could not decide between us who to recommend. He suggested tossing a coin to settle the issue but I politely declined and said that "I am much younger than Mr. Tabamo and have more opportunities for advancement later". I suggested that Mr. Arafiles choose Mr. Tabamo. Upon saying that, Mr. Tabamo shook my hand in appreciation of my gesture. Mr. Arafiles chose me to be his AWSC much later. Mr. Tabamo returned to the Climatology and Agrometeorology Division.

I graduated with a Master's degree in Meteorology on April 26, 1981. I resigned from Feati U, after having been promoted to AWSC, NIAS, in December 1982. I succeeded Atty. Garcia and my workload as an AWSC became much heavier because Tatang Arafiles required my review on every document that he had to endorse or sign for approval. Besides assisting the Weather Services Chief, I acted as the co-ordinator of the WMO Regional Meteorological Training Centre (RMTC) for the South-West Pacific. As such, I also supervised the WMO fellows who were taking graduate courses at the DMO, UP Diliman and saw to it that the training courses being conducted at the PAGASA were according to the WMO Guidelines for the Training of Personnel in Meteorology and Operational Hydrology (WMO-258).

In 1983, I consolidated the fellowship and scholarship guidelines of the PAGASA into a memorandum circular that prescribed, among others, the responsibilities, obligations and privileges of the fellow/scholar and penalties. The guidelines included, for the first time, the stipends and allowances of a fellow/scholar taking a doctoral program. Consequently, I also prepared the guidelines for the evaluation and monitoring of researches and the award of monetary incentives for completed researches. The progress of approved researches was regularly presented in in-house seminars organized by the committee that implemented the guidelines, named Research Evaluation, Monitoring and Incentive Award (REMIA) Committee. The incentive award was 15% of the author's and his co-author's current salary and may be given to up to three research papers.

The first international symposium I attended was the WMO Symposium on Typhoons, held in Shanghai, China on October 6 to 10, 1980. China was still in the so-called "Bamboo Curtain", a Cold War euphemism for the demarcation between the Communist and capitalist states in East Asia, particularly the People's Republic of China (military.wikia.org/wiki/Bamboo Curtain). I was with Mr. *Jesus F. Flores* and Dr. *Rodolfo de Guzman*. Mr. Flores was the Chief of the Weather Division then. We had to stop for two nights in Hong Kong (HK), then a British colony since 1842, before proceeding to Shanghai. We checked in a hotel on Nathan Road, the main thoroughfare in HK. We were met late in the afternoon by some of the staff of the Hong Kong Observatory, the weather bureau of HK. They treated us to a traditional Chinese dinner, where many dishes were served one by one, first hot tea and soup, followed by four or five more dishes, before serving the last dish which is white rice.

The trip to Shanghai involved taking a train to an airport just outside the border between HK and China. There were only a few people in the Chinese train station, who were probably soldiers as they were in military uniform. We took a propeller plane that looked like a civilian aircraft to Shanghai, where we met *U Tin Win* (Burma) and Mr. *Patipat Vivatsiri* (Thailand), two of our former WMO fellows who earlier took their master's degree in UP Diliman. Eventually, Mr. Vivatsiri became the first WMO fellow to graduate with a Ph.D. from UP. U Tin Win got his later and returned to a turbulent country. Shanghai was an old city then, teeming with hundreds, maybe thousands, of people riding on bicycles during the day. There were also hundreds in the shops and other buildings, going about their business, like ants. There were only a few motor vehicles then, mostly owned by the government, which had only its parking lights on when traveling at night. I found out during our stay in this city that "lumpiang Shanghai", which was small, was just our own invention. The meat rolls in Shanghai were large, bigger than a spring roll. On our way back to HK, we made a side trip to Guangzhou, also known as Canton, but I did not find a "pansit Canton" there.

My baptism of fire came when, as a fresh master's graduate, I was nominated by the WMO to participate in the WMO/JMS/AMS Regional Scientific Conference on Tropical Meteorology, in Tsukuba, Japan, on 18 - 22 October 1982. JMS stood for Japan Meteorological Society while AMS was for American Meteorological Society. I was designated as a resource speaker in one session and Session 16 chairman. I was fortunate to be with Drs. *Mariano A. Estoque* and *Rodolfo A. de Guzman*, and Atty. *Crisostomo C. Reyes* in the conference. Dr. Estoque gave me some useful pointers on being a session chairman. He had a long experience in the U.S. in teaching meteorology, oceanography and atmospheric sciences, including climatology and hydrology, in various universities there and abroad, and most probably in attending and being a chairman in scientific conferences. He had quite a long list of published research works before he decided to settle back in the country of his birth. He later became my professor in UP, mentor and PhD dissertation reader. He was instrumental in the graduation of at least 50 MS and 20 PhD students in UP Diliman during his tenure there in the late '70s to '90s. He was the second Filipino next to Dr. *Roman L. Kintanar* to receive the International Meteorological Organization (IMO) Prize, the highest award given annually by the WMO for outstanding contributions in the field of meteorology and, since 1971, the field of operational hydrology. The Prize carried a US$10,000 bonus. The IMO was the forerunner of the WMO.

Tatang Arafiles was the president of the Philippine Meteorological Society (PMS) when it organized the Symposium on Tropical Cyclones in the Western North Pacific Ocean at the Sulo Hotel in Quezon City on November 20-23, 1982. He invited meteorologists from the PMS counterparts of Taiwan and China. He told me that he wanted them to become more friendly toward each other since some of them had relatives on both sides. He designated me as the Documentation Chairman. As such, I had to take note of the proceedings, including the discussions after each presentation of a paper of some of the participants. I prepared the proceedings after the symposium with Atty. Crisostomo C. Reyes, who was also an engineer, as the editor. Being lower in rank, I was designated only as the Associate Editor. I used the first personal computer we had in the PAGASA, utilizing Word Star. The PMS published the proceedings, which we printed at the Printing Unit of NIAGAS. This symposium was followed by three others organized by the PMS. I was already an assistant weather services chief (AWSC) when the 2nd and 3rd PMS Symposium on Tropical Cyclones in the Western North Pacific Ocean in 1984 and 1986, respectively, and the AMS/PMS Joint Conference on Applied Meteorology and Climatology in 1985 were held. I was named the documentation officer and editor of the proceedings of these symposia and conference.

People Power Revolution of 1986

The People Power Revolution (also known as EDSA Revolution, the Philippine Revolution of 1986 or the Yellow Revolution) occurred when DND Secretary Juan Ponce Enrile and Armed Forces of the Philippines Vice Chief of Staff General Fidel V. Ramos declared their withdrawal of support to President Ferdinand E. Marcos on February 22, 1986 in Camp Aguinaldo, Quezon City. Tense moments followed when thousands of their supporters responded to the call of the Archbishop of Manila, Jaime Cardinal Sin, and gathered in front of the camp and Camp Crame on EDSA. Soldiers in tanks and helicopters were just awaiting orders from Malacañang palace, where Marcos was monitoring the unfolding events which were shown on tv, to storm the Camp Aguinaldo gates. People from all walks of life, including nuns, stood in front of the tanks and offered the soldiers flowers. This scene became the icon of the EDSA People Power Revolt. The relatively bloodless revolution gained worldwide prominence and emulated by at least two countries later. Sporadic skirmishes between the administration soldiers and the Reform the Armed Forces Movement (RAM) soldiers, who backed the "rebels", took place during the following day, involving taking control of radio and television stations.

I did not know when I went to our office the following Monday, February 25, 1986, that some soldiers loyal to the administration have stationed themselves on the platform of the tv antenna beside the QCDB building. I had to park my car at the rotunda on South and West Avenues because Quezon Avenue was barricaded there. I walked to our office which was about 150 meters from the rotunda, unaware of two soldiers on the antenna platform. Some people tried to warn me not to walk across the vacant lot just before the building. I thought then that they would not probably shoot me since I was in our long sleeve barong uniform and carrying a brief case. I safely made it to my office on the Penthouse and did my usual work since there was no announcement of the suspension of office. Tatang Arafiles, who sleeps in a room adjacent to the WBECCI canteen, came to inform me about the suspension. We went down and joined the people milling in front of the building. A tv crew from Japan asked if they could go upstairs to take footages of the soldiers on the platform and Tatang hesitantly agreed. I accompanied the crew to my office and peeked through the blinds to see if the soldiers were looking in my direction. When I saw that they were not, I allowed the cameraman to train his camera to the tv antenna, warning him to be careful not to be seen by the soldiers. We left my office after they have taken a few footages. As I again came out of the building, a group of the rebel soldiers with red bands wrapped around their head arrived and went up the office. Shots were exchanged between the two groups as a black helicopter hovered above. The people caught in the situation scampered all around for fear that the helicopter may fire at them. I joined a group that took shelter at the gasoline station opposite our office, then I ran toward Times St. and West Ave. until I got to my parked car at the rotunda and sped home to safety. I saw on tv later the dead soldiers on the platform being lowered with a rope from the antenna platform. It turned out that the helicopter hovering above the soldiers at the platform belonged to the rebel group. When we returned to our office the following morning, Tatang called me to show two bullet marks and nearly shattered glass panes and blinds in his office.

Following the reestablishment of the democratic government after the ouster of Ferdinand Marcos in 1986, President Corazon C. Aquino ordered the reorganization of the National Science and Technology Authority (now called Department of Science and Technology) and all agencies under its authority, pursuant to Executive Order 128, s. 1987. Five major branches (Weather, Flood Forecasting, Climatology & Agrometeorology, Astronomical, Geophysical & Space Sciences, and National Disaster Reduction) and three support divisions (Administrative, Finance & Management, and Engineering & Maintenance) now constitute PAGASA. This organizational structure remained until October 2008, when the agency underwent a Rationalization Program pursuant to Executive Order 366, s. 2004 issued by President Gloria Macapagal-Arroyo. The Rationalization Program of the government was aimed at making the government focus its efforts on vital/core functions and enhance effectiveness and efficiency of public service (https://en.wikipedia.org/wiki/PAGASA). It was then that the NIAGAS was renamed as the Atmospheric, Geophysical and Space Sciences Branch (AGSSB).

I have become so engrossed in my career in the late '70s and early '80s that I cannot now remember the important events in our children's childhood. I deeply regretted this now! All I can recall is that I was an over-protective father, who was strict in disciplining his children so that they will grow up to be good adults. The use of marijuana and other similar depressants/stimulants was starting to increase in our neighborhood and I sternly warned my sons to stay away from these vices. I told them I would personally bring them to the police if I smelled the "grass" from them.

Dramatic scenes during the EDSA People Power Revolution in 1986.

I remembered, however, that Connie and I saw them through to complete their entire education. Connie and I attended all their school events, particularly their presentations, Mich's electronic organ recital, and graduation ceremonies. I also remembered an instance when our car lost its brakes while we were on our way in Proj. 8 from the SMC, after turning from Congressional Avenue. I was with our three children and the three Olona children then. I initially felt nervous but kept my presence of mind by putting the car on second gear, driving very slowly and using the hand brake when it was necessary to stop the vehicle. Thank God, we reached home safely.

I cannot forget, too, how excited our two sons were when we brought them, together with their sister, to the SM Mall in Cubao, QC during the late '70s. They rushed to the display cabinets of Voltes V and other Japanese robots as soon as we arrived at the mall. Voltes V was a Japanese anime tv series that first aired on April 6, 1977. It became a hit in the Philippines and our sons, including me, liked to watch it on tv. We even sang almost perfectly its theme song in Japanese when the show started and ended. Being their doting parents, we bought them both the smaller robots, which were within our financial reach. They both had a collection of the robots, including Mazinger Z. Mich was fond of Hello Kitty products, many of which we bought for her, too. She was also very happy every time I brought home chocolates, especially when I attended meetings at the World Meteorological Organization headquarters in Geneva, Switzerland. I usually bought three kilos of Swiss chocolates, mostly for her. This may be the reason for her love of this confectionery today.

All our children finished elementary school at SMC. Mich received most of my attention in tutoring in the evening, when I had no classes. She continued her high school in SMC, while Chris enrolled and completed high school at La Salle Greenhills in San Juan City. He then enrolled at the De La Salle University on Taft Ave. to pursue his B.S. in Computer Engineering but transferred to the AMA Computer College (AMACC) in Project 2, QC in 1989 because of the long travel time to Manila. Mich graduated in SMC in 1987 and took a B.S. in Tourism course in UP Diliman. She had *Venancio (Benjie) Paras* as a classmate and graduated with him in 1991. Benjie Paras was later named the Most Valuable Player and Rookie of the Year in the Philippine Basketball Association in one season. Jude went to Notre Dame of Greater Manila in Caloocan City, where he graduated in 1991. He obtained the degree of Bachelor of Fine Arts, major in advertising, at the University of Santo Tomas, Manila in 1995.

When Mich was about to turn 18 in 1988, we asked her if she wanted a debut party to celebrate her birthday. She said she wanted to go to Hong Kong instead of a party. She was accompanied by her mother there and spent three days going to the tourist spots, including the famous Victoria Peak. Hong Kong's Disneyland was not existent then.

Meanwhile, I was promoted on May 4, 1989 to a Weather Services Chief (WSC) position, which became vacant when Mr. Arafiles was promoted as a deputy director. I became the chief of the AGSSB, which had 106 personnel, including myself. The AGSSB had three sections, namely; the Training Section, the Geophysical and Atmospheric Research and Development Section (GARDS), and the Astronomy Research and Development Section (AsRDS). The training section hosts the WMO Regional Training Center (RTC) for the South-West Pacific Region, the PAGASA Library, and the Scholarship and Fellowship Unit.

The GARDS has the Air-Sea Interaction Research Unit and the Instruments Development and Research Unit, which carries out meteorological instrument development, calibration and repair, and operates a wind tunnel for the calibration of wind instruments. The AsRDS has the Astronomical Observation Unit, which operates the PAGASA Observatory in UP Diliman, the Time Service Unit, the Astronomical Publication Unit, and the Planetarium Unit.

Chris got married first among our children. He was only 18 years old when he and his girlfriend, *Maricel (Cel) S. Mirabel,* 19, who lived with our neighbor and her aunt *Filipina (Philip) S. Olona,* had a relationship. Both of them kept Cel's pregnancy a secret from her aunt and us. Cel wore a girder to hide her bulging tummy as long as possible. When her stomach became so large that it had become dangerous for her child to be restricted by the girder, Chris was forced to divulge their secret to us. Upon knowing this, I told Chris that he must marry Cel as soon as possible. They were married by Bulacan Judge *Aurelio Sebastian,* who was the father of my officemate, on November 29, 1990 at the Shangri-La Restaurant, near West Avenue, QC. They also got married later in a catholic church on Del Monte Avenue, QC.

The AGSSB Family photo collage, 2005

Their first child, and our first grandchild, was a girl, born on January 21, 1991 at the Bernardino Hospital, a kilometer away from ACF Homes on the Quirino Highway. They named her *Christine Bernadette,* the female counterpart of her father's name, and gave her the nickname Vernie to distinguish her from me. Chris joined Cel in the Olona home, which was three houses from ours, after their marriage and got a part-time job at Jollibee in Novaliches town proper. He continued going to AMACC but he lost so much weight with his schooling, work and family duties that I advised him to stop working and just concentrate on his studies. We also asked him and his family to move to our home. They did so in 1992 before Vernie's first birthday. Chris went back to work for Philippine Council for NGO Certification (PCNC) in Makati City from 1996 to 1998, when their second child was born on May 8, 1997, to augment our support for his family. The child was a male, who they named *Charles (Cha) Michael*. He then worked with CableSys, which is a subcontractor of the PLDT together with PCNC, also based in Makati City, up to 2000. He was hired to use AutoCad, converting plans drawn by engineers into digital plans. Having a second child somehow delayed his schooling but he eventually graduated with the degree of B.S. in Computer Engineering from the AMACC in 2000. It took about four more years before Chris landed a job in the Celebrity Cruises, a cruise line that offers leisure trips aboard its ships with destinations to the Caribbean, Europe, Alaska, Bermuda, the Bahamas, Galapagos, Asia, Australia, Mediterranean, Northern Europe, and Greek Islands.

Vernie completed her B.S. in Hotel and Restaurant Management at the Philippine Women's University in Quezon City in 2009. She had been in a relationship with *Hanzel Allan P. Estrada* since 2007, who married her after giving birth to their first child, a female who they named *Cianna Haylee*, on March 26, 2012. Cianna is a cheerful and a smart girl, who makes us happy every time we see her antics on Facebook. She is

a consistent honor student since her kindergarten up to school year 2018-2019 when she was a second honor student in her school in Rocka Ville, Plaridel City, Bulacan. We always enjoy and look forward to seeing her on Facebook. Their second child, a male, was born on Aug. 1, 2016 and named *Cendric Hernest*, nicknamed Drake. He is a cute little boy, who is also a cheerful and a smart boy like his ate Cianna. I hope he will also be a consistent honor student because of his large head like myself. Photo at right

shows Drake and Cianna with me when we last visited the Philippines from Nov. 2017 to May 2018.

Mich got employed at the PLDT in 1991 and was assigned at the Valenzuela service center while her mother has been transferred to the Sampaloc Service Center near the Welcome (now Mabuhay) Rotunda, the boundary between Manila and Quezon City. She met there and fell in love with *Rolando (Roland) Esquejo Carpio*, who would become her husband on January 8, 1994. He was and still is a very industrious man, who is a jack-of-all-trades. He can be a mechanic, carpenter, tailor, barber and a cook. They lived together in a small house in Mandaluyong City, near the relatives of Roland. Shortly later, we invited them to use one of the two rooms, which we had added to our original two-bedroom house. We used the bigger additional room while they occupied the smaller one which was bigger than the original ones. Chris and his family stayed in our former room, while Jude shared the last room with Nang.

Mich delivered her first child, a female, at the Chinese General Hospital in Blumentritt, Manila in 1994. They named her *Rochele Carla* (the first name is a combination of the first and last syllables of her parents' name). She had a second child, a male, two years later. They named him *Roland Miro*. Before Connie chose the optional retirement offered by the PLDT with attractive benefits in 1998, she made a loan from the SSS to buy a house and lot in Rocka Ville, Plaridel, Bulacan. The house was completed in 1997 and we rented the house to a family, who was a friend of the family opposite the house. The lessee was at first on time with the monthly rental payments which became irregular as the months went by. When the payment did not come for three months, we went to the house and found only a man who was taking care of the house. We decided to let him leave with a warning that we will call the sheriff's office if he did not leave. Mich and her family transferred to Rocka Ville in 1998. Connie retired on December 31, 1998 and the cost of the house and lot was deducted from her retirement benefits.

In 2001, Connie and I encouraged our children to apply for immigration to Canada. I was inspired to do this after learning that some of my officemates have been approved for immigration to that country through the Canadian Immigration Consultancy (CIC) in Pasay City. We all went and took the evaluation process conducted by the CIC to determine whether we are qualified to go to Canada as working immigrants. All our children qualified but only Roland was determined to proceed as soon as possible. We paid the processing fee for his immigration and he finally left for Canada in 2003. Mich and her two kids followed in 2004 after Roland found a job there. They lived in a ground floor apartment on Victoria Park Avenue, Scarborough, Ontario, Canada.

Jude met and fell in love with *Maria Felisa* (*Marife*) *Pedreña*, who was already a dentist when they met. Marife had a fair skin and long hair, traits that Jude was inclined to like in women. His former girlfriend had the same features. Jude and Marife got married in 2000 and had their first daughter, who they named *Angel Beatrice*, in November 2000. They lived together in Marife's grandmother's apartments in San Francisco del Monte. Later, she studied orthodontics, which we supported, and lent them an initial capital to open a dental clinic at the corner of Quezon Blvd. Ext. and Roosevelt Ave. atop the Mercury Drug Store. She had many patients, including call center agents, sales representatives and receptionists from nearby offices. She earned so much from the clinic and her relationship with Jude suffered and started to diminish. In December 2009, Jude's marriage with Marife has become difficult after their second daughter, *Blessie Marie*, was born in October 2009. Jude then looked famished due to his relationship problem. In April 2010, Jude and his wife separated. He filed for legal separation in 2013 and was granted a "void ab initio" decision in 2016.

Meanwhile, Jude met and got in a relationship with *Arlyn* (*Ara*) *C. Italio*, in 2012. Ara is a native of Sara, Iloilo, and worked as a receptionist for an agency that deals with Filipino overseas maritime workers. She used to be a member of a crew aboard a cruise ship. Jude and Ara got married in a civil ceremony in Quezon City on May 18, 2018 (Photos next page), almost two years after Jude's legal separation with his former wife was approved. Ara's parents, *Ariel Italio* and *Evelyn Casamayor*, and relatives attended the ceremony, while Chris and his wife and granddaughter, Cianna, also came with us. Chris and Maricel were the principal witnesses as well as Ara's aunt *Anelyn Casamayor Veloso* and brother *Arnel* (*Jimboy*) *C. Italio*. Jimboy was with his daughter, *Xianelle*. We all enjoyed the reception lunch at the Shangri-La Restaurant on Times St., QC.

Ara went back to Iloilo City a week after their wedding to finish her course in B.S. in Cruise Ship Management at the John B. Lacson Foundation Maritime University. She graduated in April 2019 and is now back to work as a front desk agent in an apartelle in Mandaluyong City. We are so happy and grateful that Jo met and married a very beautiful and good woman, who captured my heart, too, when I first met her. This is one reason that I have encouraged them both to apply for immigration here in Canada as soon as possible.

Introduction to Space Science and onwards

I was introduced to the International Astronautical Federation (IAF) when I received an invitation to attend its UN/IAF Workshop on Organizing Space Activities in Developing Countries: Resources and Mechanisms and 44th IAF Congress, in Graz, Austria, on 15-22 October 1993. I must admit that I was completely ignorant of the Federation and its activities and I decided to apply to attend the Workshop with the full support it offered. I now think that it was aware that, being a developing country, the Philippines had just embarked to get into space science in 1988 and the IAF remained true to its mission of international space advocacy to establish a dialogue between scientists around the world and to lay the foundation for international space cooperation. In that last aspect, I think that it was successful in implanting its "seed" in my mind that inspired me to have a keen interest in space science later.

Connie had her first trip abroad when she joined me in Jerusalem, Israel, where I attended the UN/IAF International Symposium on Benefits of Space Technology for the Developing World - From Economic Growth to Environmental Protection, 6-9 October 1994 and the 45th International Astronautical Federation Congress, 10-14 October 1994. She followed me on October 6 because we were unable to obtain a plane booking together the day before. I left the symposium during the opening day coffee break to fetch Connie in Tel Aviv, where the international airport was located. It was the first time I took a taxi with many co-passengers to and from the airport. Connie was happy to see me at the exit gate and later told me of her tense experiences at the Manila and Bangkok airports. Being alone, she was asked by the Bureau of Immigration officials (she was told to go to the airport immigration office) where she was going. When she told the officials about her purpose of joining me in Israel, they seemed doubtful of her answer and asked if she was employed. They even called and spoke with Connie's personnel manager to confirm her employment. It was apparent that they were trying to extract money from her to facilitate her passage through immigration, but Connie did not oblige them and they had to let her through in time to catch her plane. In Bangkok, she encountered the same young people, who interrogated me when I our plane stopped there. They were Israeli security personnel, who asked relevant questions. My checked-in luggage was already present when they asked me if it belonged to me and whether I left it unattended in Manila. They also asked if I had accepted any package from somebody for delivery to my destination. My answers were positive to the first question and negative to both the other questions. They asked me to open my luggage and visually inspected its contents. They let me through the transfer gate to El Al, the Israeli airline, after they were satisfied with the inspection.

The afternoon of the last day of the Symposium was left open for the participants to see Israel. Connie and I joined a group to the Dead Sea and the Masada. I took a dip in the Dead Sea, which has the lowest elevation and is the lowest body of water on Earth. It was given that name because no fish can survive in its salty waters. As advertised, I merely floated and did not sink when I laid out and stretched my arms in it. Its black mud is also said to be medicinal if one wiped it on the skin. I bought a pack of the mud for Connie, who is health conscious. After this, we went up the top of Masada, where on April 15, 73 A.D., 960 Judean rebels except two women and five children, who hid in the cisterns and later told their stories, took their own lives rather than live as Roman slaves (history.com/topics/ancient-middle-east/masada).

I was not a member of the IAF, so I was technically free on Oct. 10-14, 1994. I, however, looked around the venue and watched the posters in the foyer. The International Astronautical Federation (IAF) (French: Fédération internationale d'astronautique) is an international space advocacy organization based in Paris, and founded in 1951 as a non-governmental organization to establish a dialogue between scientists around the world and to lay the information for international space cooperation (Wikipedia). The posters roused my interest in space science of which I had no previous knowledge before I joined the Weather Bureau. I learned that it did not only involve space exploration but also the development of technologies that enabled humans to go farther beyond our atmosphere. One is the miniaturization technology that allowed the manufacture of high-speed computing devices in space modules and other applications like cellular phones, communications, global positioning system and meteorological satellites, and the construction of space telescopes like the Hubble Space Telescope. Another technology related to this is the low-level satellites and earth-looking telescopes for natural hazards and resources monitoring, which are greatly beneficial to countries like the Philippines, that are frequently affected by cyclones/hurricanes, intense precipitations, floods, earthquakes and volcanic eruptions.

I took the rest of the days in Israel to explore as a Catholic the Holy Land with Connie. We joined a tour around Jerusalem, walking the Way of the Cross or Via Dolorosa, stopping at the Mount of Calvary, where Jesus was crucified, and the Holy Sepulchre where He was entombed. I entered the Sepulchre, which was a small room that can accommodate only two to three adults, and knelt down to pray beside the stone bed where Jesus was laid down. I had a cold feeling when I prayed, which gave me goose bumps. Mount Calvary and the Holy Sepulchre were now housed in a church named after the latter. We likewise walked around the city, which was lined by houses of similar ochre color, mostly with solar panels and satellite antennas on its roof. I felt then like I was walking along the streets where Jesus walked, too, and felt serenity in my soul.

We next joined another tour by bus to Bethlehem, the birthplace of Jesus, just 5 kms south of Jerusalem, in the West Bank. I did not know then that the place was under the control of the Palestinian National Authority. This required visitors to present their passport when they entered the city. Connie and I saw the exact spot on which Jesus was born. I took a photo of her touching the spot, now decorated with the golden rays of the Star of Bethlehem around it.

My trip together with Connie to Jerusalem had changed my outlook in life without my realization then. It deepened my belief in my religion and initiated my great interest in space science. At present, I think space science is the key to our future as humanity. I believe now that this was the start of my destiny.

Connie touches the spot where Jesus was laid on the lap of Mary from the cross.

Masada, Israel - 9 Oct. 1994

Connie before boarding with me the cable car to Masada.

Prior to 1994, I attended numerous (22) local and international training seminars, workshops, symposia, conferences and meetings. I have been to all the places I dreamt of seeing when I was young, except two, notably Egypt for its Pyramids in Giza and Argentina for the Christ the Redeemer statue in Buenos Aires. I went to Japan (3), Thailand, India, Malaysia, Switzerland (2), US (2), and Austria from 1973 to 1993. From 1994 to 2008, I went to Indonesia, Israel (2), Thailand, Brunei Darussalam, Sri Lanka, Switzerland (2), Singapore, France, Russian Federation, Australia, Japan, UK, South Korea, Spain and Turkey for a total of 29 more.

All of these were related to my administrative and scientific/technical work, except space science which was introduced with the establishment of the PAGASA in 1972. The only aspect of the agency's functions that was closely related to space science was its astronomical services. However, the services were confined to observations of sunspots, lunar occultations, the satellites of Jupiter and transits of Mercury, comets and other planets, using its 29-inch reflector telescope. Others services included its time service, calendar production, and planetarium shows. These limited activities may have been the principal reason that the Astronomy Research and Development Section (AsRDS), which was transferred to the AGSSB, has not been elevated to a Division. The staff of the AsRDS, majority of whom were undergraduates, do not have a formal education in astronomy. The knowledge in astronomy that they possessed was obtained through the infrequent in-service training courses conducted by the agency and through the books that were procured, usually from overseas sources. Other than the 12 small telescopes, the additional digital 45-cm telescope, an 88-seat fixed planetarium, a mobile planetarium and its new Precise Time Scale System that was designed to establish a national timing reference and to satisfy the requirement of the existing laws on time and frequency services in the country, the AsRDS had a few resources and outputs to justify its promotion to a division. Its products are: (dissemination of the) Philippine Standard Time; promotion of Astronomy through stargazing/telescoping sessions and planetarium shows; and planetarium tour in selected areas in Luzon (http://bagong.pagasa.dost.gov.ph/ products-and-services). However, the passage of the Philippine Space Agency (PhilSA) Act into law by President *Rodrigo R. Duterte* on August 8, 2019 provided hope for the AsRDS to rise to a Division, if not a separate agency within the PhilSA.

Tatang Arafiles retired on his 65th birthday on April 9, 1991. Being without his family in the Philippines, who had immigrated to the US, except his son, Dr. *Ruben P. Arafiles*, he lived in a small hut built as a storage building in the Science Garden. Dr. Arafiles, who was a noted orthopedic surgeon, lived in a subdivision in Parañaque City with his family. His wife was an anesthesiologist at the Asian Hospital and Medical Center in Muntinlupa, Rizal. Tatang was a proud man, who did not want to impose on his family, particularly his son. He chose to live alone and looked miserable at times because of self-negligence or old age.

I continued to revere and respect my predecessor by awarding him a plaque of appreciation as a former chairman of the board of directors (BOD) of the Weather Bureau Credit Cooperative, Inc. (WBCCI) when I was its chairman in 1999. I led the WBCCI in requesting the PAGASA director to complete for the Cooperative's use the unfinished building, meant to house the radio station DZCA of the agency. The radio station was in a room at the Planetarium, beside the building at the Science Garden. We completed and inaugurated the building in 2005, the last year my second term as chairman of the Cooperative. I must give due credit to my close friend *Rolando* (*Totek*) *G. Valenzuela,* who led the completion of the building as one of the Board members. Totek and I became good friends after I heard him giving an impressive lecture at the Whiterock Resort Hotel in Subic, Zambales. We had a heated argument about the propriety of his editorial in the WBCCI newsletter, criticizing the sending of five BOD members to a seminar in Bacolod City, whose travel expenses were charged against the educational fund of the Cooperative. We remained good friends after the incident.

We shared the building with our sister cooperative, the PAGASA Employees Consumers Cooperative, Inc. (PECCI), and the Philippine Meteorological Society (PMS). The two cooperatives were merged after my retirement in 2008 into the Weather Bureau Multi-Purpose Cooperative (WBMPC) with our former WBCCI secretary, Ms. *Elenita S.P. Ingalla,* as the General Manager.

Natural Disasters in the Philippines

The Philippines is prone to natural hazards and/or natural disasters, being on the so-called "Pacific Ring of Fire". Natural hazards refer to all atmospheric, hydrologic, geologic (especially seismic and volcanic), and wildfire phenomena that, because of their location, severity, and frequency, have the potential to affect humans, their structures, or their activities adversely. The qualifier "natural" eliminates such exclusively manmade phenomena as war, pollution, and chemical contamination. Hazards to human beings not necessarily related to the physical environment, such as infectious disease, are also excluded from consideration here (oas.org/dsd/publications/Unit/oea54e/ch05.htm). A natural disaster is a major adverse event resulting from natural processes of the Earth. Examples are floods, hurricanes, tornadoes, volcanic eruptions, earthquakes, tsunamis, and other geologic processes (en.wikipedia.org/wiki/ Natural_disaster). According to Oxford Dictionaries, a disaster is a sudden event, such as an accident or a natural catastrophe, that causes great damage or loss of life. Six other references define the term in similar fashion, including en.wikipedia.org. In view of these definitions, I often wondered why many people and organizations, like Wikipedia and the USGS, called natural phenomena such as those named above as natural disaster, even if they do not cause great damage or loss of life, especially in unpopulated or sparsely populated areas like the oceans or seas.

The Pacific Ring of Fire is the name that is given to a horseshoe shaped area (See photo below) in the Pacific Ocean which extends from South America and North America to Eastern Asia, Australia and New Zealand. This area is famous for its constant seismic activity and because of the amount (sic) of active volcanoes that can be found here. 75% of dormant and active volcanoes are found in the Pacific Ring of Fire (www.basicplanet.com/pacific-ring-fire).

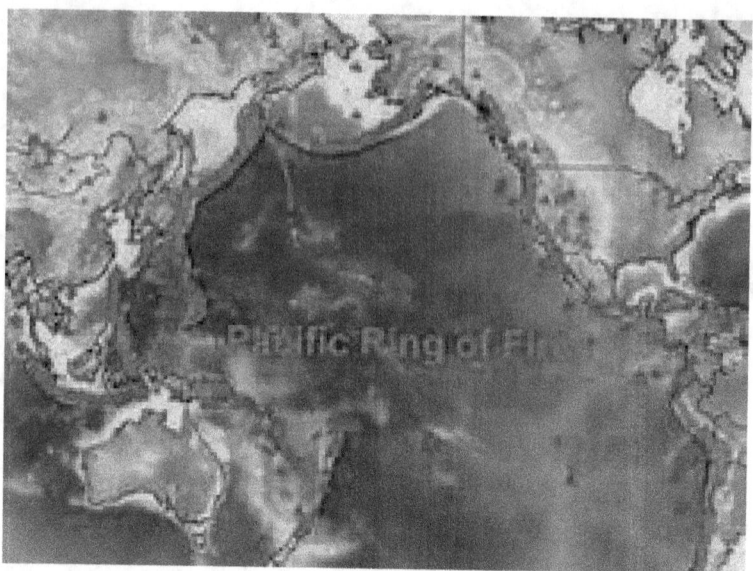

Big Luzon Quake, 1990

It was past 4 PM on Monday, July 16, 1990, when we felt the QCDB Bldg. shake at first vertically, then horizontally. We felt the shaking more strongly since we were on the top (11th) floor of the building. My officemates began to panic when the shaking lasted long that even I thought that it was our end. Two of my female colleagues (*Shirley V. Almazan* and *Fredolina D. Baldonado*), whose pregnancy was so advanced that they looked that they had already a basketball-sized stomach, embraced me while crying for fear of their life and their unborn child. I told them to stay calm and brought them beside a square concrete post of my office. We stood there until the shaking, which lasted for about a minute, stopped and I immediately told all my colleagues on the floor to vacate the building on foot, being aware of the aftershocks that will follow the big tremor. Before going down the stairs, I inspected all the five rooms on the floor. I looked at the EDP Section at the opposite end of the floor, then the PAGASA Library and finally the office of the AWSC for research where I found two female colleagues (*Aldina D. Miranda* and *Aniceta B. Garcia*) nervously cringing in fear and crying under a desk. I told them to run downstairs immediately. When we were on the ground, I dismissed all of my co-employees and told them to go home since it was already near 5 PM.

The evening news on tv reported that the earthquake had a 7.7 magnitude and its epicenter was in Casiguran, Aurora, near Cabanatuan City. Reports said that the quake killed over 1,600 people and injured hundreds of others. The reports also showed the extent of damages in Baguio City, where a hotel fell on its side and a public market collapsed like the Ruby Tower in Manila in 1968. Many other structures were completely damaged, most notably in Cabanatuan City, where the Christian College of the Philippines collapsed and killed about 154 people.

Baguio City hotel

Casualties in the Christian College of the Philippines, Cabanatuan City.

Mt. Pinatubo Eruption, 1991

One of the volcanoes in the Philippines that have remained dormant for many years is Mt. Pinatubo, located at the intersection of the borders of Zambales, Tarlac and Pampanga provinces. It is situated on the Cabusilan Mountain range separating the west coast of Luzon. The volcano has slept for more than 600 years before it awakened on June 15, 1991. It erupted and spewed gas-charged magma that exploded into umbrella ash clouds, hot flows of gas and ash descended down its flanks, and lahars swept down valleys. It was the largest volcanic eruption in the last 100 years, according to the U.S. Geological Survey (USGS). However, according to https://www.livescience.com, it was only the second, next only to a 1912 eruption of Novarupta on the sparsely populated Alaskan Peninsula. Pinatubo's eruption was, however, considered the most notorious for its Volcanic Explosivity Index (VEI).

Pinatubo's eruption occurred as Typhoon Yunya was crossing Luzon. The column of ash spewed by the volcano rose almost 34 kilometers into the atmosphere and opened like an umbrella to form a cloud 400 km across. Pinatubo's ash mixed with Yunya's rain on its way down, creating concrete-like mud that collapsed roofs more than 15 kms away. When the climactic eruptions subsided, a snow-like blanket of ash coated 7,500 sq kms of the island of Luzon. Over the next year, the ash spewed into the atmosphere by Pinatubo lowered global temperatures by 0.5 degree Celsius. Only a few hundred people perished in the initial eruption, thanks to round-the-clock volcano monitoring and proactive evacuation plans. Fine ash fell as far away as the Indian Ocean, and satellites tracked the ash cloud several times around the globe.

The Philippine Institute of Volcanology and Seismology (PHIVOLCS), led by Dr. *Raymundo S. Punongbayan*, had a major role in monitoring the volcano since loud rumbling noises were heard around Pinatubo on August 3, 1990, about two weeks after the magnitude 7.8 earthquake occurred 100 km northeast of the volcano. Pinatubo continued smoldering and smoking for months. And for years afterward, the monsoon season rains washed the ash down the mountain slopes in devastating lahars and raised the initial death toll of the eruption from 200 or 300 to more than 700. However, although the eruption was the most notorious and second largest in the last century, its death toll was remarkably small. At least 20,000 lives were saved due to aggressive monitoring and public information campaigns principally by PHIVOLCS. I personally saw the wastelands in Pampanga that the lahar caused months after the eruption. Photo at right shows the tall ash cloud from Pinatubo on June 15, 1991 (Photo above courtesy of PHIVOLCS).

First PhD in the family

I became the first and only holder of a doctor of philosophy (PhD) degree in both Monillas and Soriano families on April 26, 1991, when I graduated with that degree from the University of the Philippines in Diliman, QC. Earning the degree was an arduous task that I had to endure for six long years while being a family man, an employee and a student, all at the same time. The university rules after 1987 allowed only five years for a student to complete a doctoral program. However, I began my program in 1985 and was not affected by the new rules. I was extremely happy when I graduated with Connie and Daddy as my witnesses, together with my fellow graduates, Drs. *San Hla Thaw* (Myanmar), *Paulus Agus Winarso* (Indonesia) and *Nguyen Thi Thanh Tai* (Vietnam), who were all WMO fellows.

I first worked in 1987 on how to model numerically the interaction between the sea surface and the atmosphere at the onset of a cyclogenesis. I learned in the university that there are six requirements for tropical cyclogenesis: (1) a sufficiently warm (at least, 26.5 degrees Celsius) sea surface; (2) atmospheric instability; (3) high humidity in the lower to middle levels of the troposphere; (4) enough Coriolis force to sustain a low-pressure center; (5) a pre-existing low-level disturbance; and (6) low vertical wind shear. These are highly technical terms or conditions that are difficult to understand if one has no physics or science background. I would advise readers to "google" these terms to gain more understanding on the requirements for cyclogenesis.

I had been working on my dissertation for one year when Dr. *Mariano A. Estoque*, who was my professor at the DMO, UP Diliman, casually asked me one day how my work was proceeding. I told him that I faced a wall that is hard to climb. He advised me to use his model of the land-sea breeze circulation to simulate the effect of small disturbances on the general circulation. He gave me a copy of the complete computer program of his model, which was written in Fortran. I carefully inputted the numerical model into the VAX computer at the Typhoon Moderation Research and Development Office (TMRDO) of the PAGASA, which is beside the Weather and Flood Forecasting Center (WFFC) on Agham Road, QC, opposite the Science Garden. The VAX computer was relatively slow that I had to wait for more than one hour to get an output. I went to the DMO, which then had several new IBM 385s, to run my program but I had to convert it to a Basic computer language to have an output faster. This took me much time to learn again but eventually succeeded in converting Dr. Estoque's numerical model to Basic. The model ran just about five minutes on the IBM 385, using floppy disks with limited storage capacity. Each 3.5-inch floppy disk could only contain 720 kilobytes (KB) of data. I designed several experiments by setting various initial conditions and ran the model until I completed my experiments in 1990.

I wrote my dissertation, entitled "Influence of environmental conditions on tropical waves" in only two months, defended it in June 1990 and passed the oral examination. The examination panel was composed of Drs. *Jorge G. de las Alas, Mariano A. Estoque, Israel D. Bentillo, Leoncio A. Amadore* and *Emmanuel G. Anglo.* Dr. de las Alas was my dissertation adviser while Dr. Estoque was my dissertation reader.

Looking back and thinking about my experience as a meteorologist in the PAGASA, I often wondered why the clouds do not have the same or similar shapes at a particular time and place. It was only recently when I came across the term "butterfly effect" that I realized the answer to my wonderment. The term is defined by merriam-webster.com as "a property of chaotic systems (such as the atmosphere) by which small changes in initial conditions can lead to large-scale and unpredictable variation in the future state of the system". Knowing the effects of the environment - such as terrain, geographic location, wind systems, bodies of water, and time of the year - on the weather and climate of a particular place, I came to the conclusion that the small movements of people, animals, motor vehicles, and nature itself like rivers, seas and oceans, winds and movements of the leaves of a tree, have a profound and variable effect on the formation and shape of clouds in the atmosphere.

Total Solar Eclipse of 1995 (TSE '95)

The next significant event that took place during my career at the PAGASA (Weather Bureau) was the Total Solar Eclipse (TSE '95) that will be visible in Tawi-Tawi on October 24, 1995. Tawi-Tawi was one of the five provinces of the Autonomous Region of Muslim Mindanao (ARMM), with Basilan, Sulu, Maguindanao and Lanao del Sur as the other provinces. It is the southernmost province while the Batanes province is at the northernmost tip of the Philippine archipelago.

Preparations in the agency, mainly by the Astronomy Research and Development Section (AsRDS), AGSSB, were made as soon as the year began. We issued a press release to announce the eclipse and the plans that were made. Some of the activities we lined up were visits to coordinate with the local government concerning the eclipse and preparations that were to be undertaken at the eclipse site in Languyan, where eclipse totality was to pass.

I led a team of three to see the governor of Tawi-Tawi, *Gerry Matba,* in February 1995 to solicit his assistance in the clearing of suitable spots we identified on a map in Languyan to position our telescopes to observe and photograph the eclipse. We left quite early in the morning to catch the first plane to Zamboanga City, where we took a connecting flight to Bongao, the capital of Tawi-Tawi. We were met at the airport by some of his staff, who were armed with hand guns and rifles and who drove us to a hotel owned by the governor. After the introductions and exchanges of pleasantries, we informed him of the important information about the eclipse and told him that this will be a great opportunity for his province to be known nationwide, if not worldwide, since the last total eclipse in Davao in March 1988 resulted to the improvement of the economic status of the province. The governor was glad to hear this and volunteered to mobilize the province's resources to prepare the observation posts we earlier identified. He likewise told us that he will construct temporary housing in the sites for our personnel and other visitors during the eclipse. We were surprised after the meeting that the governor has prepared a long table luncheon for us with plenty of sea foods, particularly sliced giant clam soaked in vinegar with lots of sliced onions, red hot pepper and black pepper, roasted octopi and squids, lobsters, etc.

Back at the PAGASA central office, we drafted a Presidential Proclamation for the creation of a national committee for the observation of the TSE '95. The committee will be chaired by the DOST and its members are the Departments of National Defense, Tourism, and Transportation and Telecommunications, and PAGASA. Monthly meetings were held at the PAGASA to monitor the progress of the preparations for the event. Members of the media (press, radio and tv) and astronomical societies were invited to witness and record the event. We invited the People's Radio and Television Network and the ABS-CBN to send their teams to beam the eclipse to their nationwide audience. We also invited members of the Philippine Astronomical Society, the UP Astronomical Society and the Astronomical League of the Philippines (ALP) as well as the National Institute of Geological Sciences (NIGS), UP Diliman to send their contingents. They will be transported to and from Languyan by a Philippine navy ship and housed in temporary shelters on the observation site and the nearby school rooms. Food supplies for their cooking will be provided by the governor's office while they are in Languyan. A commemorative marker for the eclipse will be built on the school grounds. The Philippine Postal Agency (PPA) also issued its commemorative stamp on the total solar eclipse of 1995, through the efforts of the advertising agency Ibex International Inc. of *Percival (Percy) Campoamor Cruz*. PPA sent me a copy of its first issue of the stamp.

I made two or three more trips with the AsRDS staff to Tawi-Tawi in the following months to give a briefing on the TSE '95 and other topics in astronomy and meteorology to a group of teachers in Bongao and to see the progress of preparations for the eclipse. The AGSSB conducts seminars in science, in particular meteorology and astronomy, for teachers during the summer school breaks in the Greater Manila Area and nearby provinces. In one of my trips, before dinner at his fish pond, Gov. Matba related to me that he was once a member of the Moro National Liberation Front (MNLF), before laying down his arms and worked as a school teacher. He was then sipping a cup of black soup of carabao which he shared with me as he was telling his story. He told me that there was not a day then that he did not wash his face with the blood of people who he decapitated. He said that the bullets fired in the island were more than the blades of grass on the ground. He also told me sensitive information that were detrimental to the government and which I cannot share here for security reasons. It is enough to say that my conversation with a non-commissioned officer guarding the 16-seater army plane I was about to board back to Zamboanga can confirm the governor's revelations. I asked if he was from Ilocos because I can tell from his appearance and he answered positively. When I asked further why he is there and not in Luzon, where his family is and there is less danger from rebels, he replied that he liked it there.

I was with *Elenita (Lenie) S.P. Ingalla*, Dr. *Amelia C. Ancog*, Dr. *Carina (Caring) G. Lao, Lita Suerte*, and *Joan Bondoc* on my next trip to Tawi-Tawi. Lenie was my efficient secretary at the national committee meetings and my activities related to the TSE '95. Dr. Ancog was the DOST Undersecretary with her assistant Lita while Joan was an award-winning photojournalist of the Philippine Daily Inquirer where she had worked for the last 14 years. Dr. Lao was the AGSSB AWSC. Gov. Matba personally drove his speedboat with four powerful outboard motors and four bodyguards with short modern uzis hanging from their shoulder, to Languyan for Usec. Ancog and our party. Our nervousness with the guards was amplified as he drove the speedboat at top speed even on large waves. There was nothing, except the wooden panel separating the front and rear parts of the boat, that we can hold on firmly to avert us being thrown off when it struck the big waves. See middle left photo below where Joan, Lenie and I try to hide our nervousness with a smile.

Gov. Matba brought us to his pearl farm before going to Languyan. He gave each of us a malong to change into because of our wet clothes by the waves during the speedboat ride. He also invited us to take two oysters with a pearl from a bunch as his gift (Top and middle center photos below). However, Joan found an oyster with two pearls in it and showed it to the governor, who said to find another two-pearl oyster to complete a set of earrings, necklace and ring. We all happily did as he said. It was unfortunate, though, that Caring gave hers to Lita, who kept Dr. Ancog's pearl oysters, for safekeeping until we came back to Manila. Upon return to our home offices, Caring informed me that Lita has not given her the oyster pearls. I had the pearls, about half an inch in diameter, mounted in Cubao, QC and gave it to Connie, which she still used up to this day for special occasions.

My team, composed of three AsRDS staff (*Elmor A. Escosia, Elenita S.P. Ingalla* and *Ruben Cunanan*) and the *Administrative* Officer (*Simeon V. Inciong*) who was with the Section prior to his promotion, was invited by the provincial government of Sulu to conduct a lecture on astronomy and announce the TSE '95 (Bottom right photo below). ARMM DOST Secretary Anni called me from Jolo to send his invitation. He was a gracious person, who calmly told me that he was a male when I called him "ma'am" because of his female-like voice. He received us at his house where we slept for a night before the seminar. It was my first time to go to Sulu, which was notorious for its separatist rebels. Connie tagged along with me as she wanted to see Jolo, too. Sec. Anni served us the sweet and delicious durian fruit for our breakfast dessert. We gave the lecture in a public school to students and teachers the following day and left for Manila in the afternoon.

I gave a press briefing on the eclipse during the last meeting of the national committee at the PAGASA Central Office, QC in the first week of October 1995.

A team of the Engineering and Technical Services Division (ETSD) of the PAGASA had gone on October 10, 1995 to the site to help in the preparations for the accommodation of the expected visitors from Manila, including about 90 officials and employees of the agency. Vice President *Joseph E. Estrada,* DOST Secretary *William G. Padolina* and PAGASA ex-Director *Roman L. Kintanar* were the important officials who will grace the significant occasion. Gov. Matba had built a special two-storey house near the site for them. I was so pleased to hear my colleagues from ETSD that Gov. Matba asked them where I was when he visited them in their lodging at the school rooms.

Three of the AsRD staff left with three of our telescopes a few days before the eclipse to set up the instruments on the eclipse totality path in Languyan (See top right photo above). Together with them were Fr. *Victor Badillo, S.J.* for whom an asteroid has been named, and the Chairman of the National Institute of Geological Sciences in UP Diliman, who set up their own instruments for the eclipse.

On October 21, 1995, three days prior to the eclipse, I left together with some of my AGSSB and AsRDS staff for Tawi-Tawi. On the following day, we held a lecture in a school for its teachers and students in Bongao on astronomy and the total solar eclipse that will take place in Languyan. We left for Languyan on a ferry boat on the night of October 22, 1995, together with many of the visitors from Manila, including some of my friends from Radio Manila, *Susana Pilar Layos* and *Melly Tenorio*. We had tense moments when the boat stopped for several long minutes in the middle of nowhere as the night was so dark that we could hardly see any star in the sky. The clouds were due to the Intertropical Convergence Zone (ITCZ) that was over the Mindanao region at the time. The ITCZ is associated with widespread cloudiness and rainfall. Susana told me later that the boat captain found after waiting for clearer skies that he was heading towards the Turtle Islands, which was on the North while our destination was on the Northeast. We safely arrived in Languyan in the early morning of October 23 and joined many others at the site.

Below is my exchange of comments with Susana Pilar Layos on FaceBook memories on October 24, 2013.

Susana: "Ay Sir, how can I forget ang isa sa pinaka-exciting at unforgettable experience ko bilang journalist when I was invited to join your group papuntang Languyan Island to witness and cover that event. Tumirik tayo sa gitna ng laot at walang katiyakan kung makakaalis pa tayo dun. I remember asking the captain of the ship "Sir, don't you have any navigational aid in this kind of situation?" Ang sagot nya: "Wala man, ga-look na lang sa stars!" (Ay Sir, how can I forget one of my most exciting and unforgettable experiences as a journalist when I was invited to join your group to witness and cover the total solar eclipse at Languyan? We stopped at the middle of the sea, without assurance that we can leave there. I remember asking the captain of the ship, "Sir, don't you have any navigational aid in this kind of situation? He answered, saying "No, I just look at the stars!")

BMS: "Oo nga (Yes indeed), Susana! That experience was truly unforgettable! God was with us then and showed the captain the right course."

On the eve of the eclipse, hundreds of residents from neighboring barangays came for the evening festivities where various groups performed their local dance presentations. A funny anecdote occurred when Dr. Kintanar lost one of his shoes in the thick mud as he walked to the school grounds where the festivities were held. Doc Kintanar's shoe was later found and returned to him. VP Estrada, who was given a pair of rubber boots, unveiled the eclipse commemorative marker in the morning (Middle right photo above) on the school grounds.

The skies were overcast with clouds on the morning of E-Day (Eclipse Day) due to the ITCZ hovering over the Mindanao Region. Together with all the people on Languyan, I was excitedly hoping and praying that a break in the overcast would permit us to see the rare total eclipse of the sun, which was computed to last more than 2 minutes over the Philippines. The large disk antennas of the People's Television Network and the ABS-CBN were ready to beam nationwide the events that will take place on Languyan soon. There were also similar setups in the cities of Baguio, Manila, Cebu and Davao ready to show the situations in their respective places.

The excitement grew as the eclipse totality approached. People were milling around the telescopes brought by the PAGASA and the three astronomical societies. One excited young muslim girl approached me and told me how fascinated she was that science can predict solar eclipses. She said she now believed in science. I told her that the occurrences of such phenomena have been calculated by many scientific organizations in the world. Schools and other groups celebrate a science month in Mindanao, no wonder the girl's interest in science.

The skies were still overcast at around 11 a.m. with no breaks visible from our vantage point. There was a remarkable reduction of sunlight and lowering of the ambient temperature by about two degrees Celsius when the eclipse totality passed over the area at Languyan. We did not see the total eclipse of the sun!

Despite this, I still considered this expedition as one of my most significant achievements during my 44-year career in the government service. I was instrumental in mobilizing massive amounts of resources (human, financial and material, and time) for this event. I sincerely hoped that I have contributed to the advancement of science in the Philippines, particularly astronomy, in promoting this spectacular astronomical phenomenon.

Tour de France et Italia

Connie and I went on a brief tour of France and Italy in July 1995, after I attended the 2nd International Conference on Computer-

Aided Learning (CAL) and Distance Learning in Meteorology in Meteo France, Toulousse, France, 24-28 July 1995. On July 27, we took a cab in Toulousse to and from Lourdes, where the Marian apparitions took place on Feb. 11 to July 16, 1858. Connie and I left by train on July 29 for Rome, the Vatican, to see the famous tourist spots there while visiting my cousin, *Grace (Chucha) S. Mariano* and her family. We visited on foot the St. Peter Basilica, the Sistine Chapel, the Trevi fountain, the Colosseum, and ruins of the old Roman senate. Of course, I took a lot of pictures of Connie. While in Rome, I arranged with our airline to re-schedule our return trip to the Philippines on Aug. 4. We took another train to go to Venice on July 31 to see the canals and the old churches there. I took more pictures of Connie there. On our return trip to Toulousse, the train stopped by Nice to unload and load passengers, as on the onward trip. We observed a group of young men with backpacks embarked and walked to and fro, apparently looking for convenient seats. They settled down at the rows behind and in front of us. As is Connie's customary behavior while sitting, she rested her feet on the bench across her. She noticed a foot under her seat, which is unusually extended. She remembered that she laid down her bag on the floor beside her and immediately picked up the bag. It was too late to call for help after she found that her purse was already gone and the men have disappeared. The purse contained the cash I shared with her and the envelope sent by my cousin to her mother in the Philippines. It was good that the thieves left her passport and other personal belongings in the bag. I informed and apologized by email to Chucha about the incident when we arrived back in Toulousse for our return trip to the Philippines via Paris. Chucha told me that the envelope had a voice tape and a letter to her mom. She did not say whether she sent money with it.

The Grotto of Lourdes

Connie by the grand canal in Venice, Italy.

Storm brewing in PAGASA

Dr. Roman L. Kintanar retired as the Director of the PAGASA on June 13, 1994, when he turned 65. He was succeeded by Dr. *Leoncio A. Amadore*, a gentle, quiet and amiable person who rose from the ranks as a weather observer in the field. He had humble beginnings in Basey, Samar. One of the projects he inherited was the installation of a surveillance radar in Basco, Batanes. Purchase of the radar was in the works, the cost of which was obtained with a congressional insertion of the gentleman from Quezon Province. The congressman, who was rumored to have a vested interest in the sale of the radar (his relatives purportedly own the importing company), was apparently pressuring higher authorities at the DOST to release the initial payment for the radar. Payment can only be made by the recipient of the item, which is the PAGASA. I saw the congressman visit our office at the Asia Trust Building, the new name of the QCDB Bldg., at least two times. I learned later that auditing rules and regulations do not allow more than 15% of the total cost of an item as an initial payment. The cost of a complete system of surveillance weather radar then was about 60 million pesos (Php 60M), which meant that PAGASA cannot pay more than nine million pesos (Php 9M) to the seller of the radar system. The congressman wanted Php 35M for the down payment.

One day in 1997, one of the assistant secretaries of DOST, Atty. *Imelda Rodriguez*, came to our office and called a meeting with the

PAGASA Executive Staff. The chief administrative officer, Ms. *Concepcion Belmonte*, and I plus several employees who were members of the Bids and Awards Committee (BAC), attended the meeting. Asec, Rodriguez seemed to be in a bad mood and angrily blurted "Nasaan na ba yan si Amadore?", as if she was looking for someone who is not a respectable person, "Ang bobo-bobo naman niyan!" We did not know why she said that and I lost my respect for her from then on.

A "rigodon de honor" in the agency ensued next. A DOST Special Order mandated that all division chiefs of the PAGASA were to be shuffled to its other divisions, like the circulation in a storm. I was assigned to the Administrative Division (AD), whose chief was sent to the Engineering and Maintenance Division (EMD), whose chief was re-assigned to AGSSB, my division. The chief of the Weather Branch (WB) was assigned to the Climatology and Agrometeorology Branch (CAB), whose chief was posted at the WB. The chiefs of the Flood Forecasting Branch (FFB) and the Natural Disaster Research and Development Branch (NDRDB) were interchanged. The chief of the Finance and Management Division (FMD) was sent to the DOST Central Office in Bicutan, Taguig, Rizal. The Deputy Director for Administration and Engineering Services of the PAGASA, served in acting capacity as chief of the FMD. The three of us who were involved in a roundabout movement, i.e. AGSSB, AD and EMD, mutually agreed to use each other's office in the morning and our respective offices in the afternoon, to sign papers and to conduct business, respectively. We all wanted to spare the efforts of moving our office furniture and personal belongings such as office decors, plaques, certificates, trophies, etc. to our new posts.

The other division chiefs complied, except the FMD chief. She went on indefinite official leave and did not report to the DOST. After several weeks, I received a memorandum from DOST, advising me to withhold the salary of the FMD chief. I consulted the Director on what action I should take regarding the memo. He asked if the person had enough leave credits, which I affirmed, then left the matter to my discretion. I did not withhold the person's salary after reading a civil service rule that allows a person to receive his/her salary even on indefinite leave, provided that that person has sufficient leave credits. The only instance when a person can be deprived of his/salary is when the person has been found to violate a civil service rule for which he/she loses the financial privilege.

The top executives of the DOST called a meeting later at the conference room of the WFFC to review the reshuffling of the officials of the PAGASA, after restoring them to their original posts. The DOST Secretary declared that only the chief of the Weather Branch complied with the DOST Special Order. I explained that the others, including myself and the chiefs of EMD and AD, complied, too. I informed the Secretary of the arrangement we made, but he did not agree that we could have had effective control of the division during our physical absence in the division. I answered that employees will continue to work even without their supervisor when they are properly motivated. I personally think that most of my colleagues were such employees. He then questioned me about my disregard for the DOST memorandum to withhold the salary of the FMD chief, who went on indefinite leave and did not report to the DOST office in Bicutan. I replied that I consulted the PAGASA director about it and I decided to allow the employee to continue to receive her salary because she had sufficient leave credits.

First sunrise of the 3nd millennium

The third one thousand years (or millennium) in the Gregorian calendar began on January 1, 2000. As early as February 1999, PAGASA through its monthly astronomical diary, made the announcement during the National Astronomy Week, which is usually celebrated in the third week of the month, that it will observe and document the 1st sunrise of the new millennium at Pusan Pt. in Davao Oriental. The announcement gained national attention after *Carlito Pablo* wrote a news item on the event for the Philippine Daily Inquirer (PDI). The local government of Davao Or. called us to inquire about the details of the first sunrise in the country. Governor *Rosalind Y. Lopez* called me by long distance telephone to express her appreciation that her province was selected for the significant event. She invited us to coordinate with her the preparations that have to be undertaken prior to it. She informed me that there is no road leading to Pusan Pt., except a motorbike path and her government will soon construct a road there. I sent a team of three to Davao Or. after talking to Gov. Lopez to make an ocular survey of Pusan Pt. and determine the best place to observe the first sunrise.

Several articles and news items were reported and published almost weekly in the PDI by Carlito, on our coordination activities with the Davao Or. government, to drum up interest about the "First Sunrise of the New Millennium" in the following months. My name frequently appeared in several broadsheets. I once had a minor debate in the press with North Cotabato Governor *Emmanuel (Manny) Piñol*, who earlier argued in an article in the PDI that Mt. Apo, with its highest peak at an elevation of 2,954 meters (9,692 feet) above sea level, should be declared the site of the first sunrise. The governor claimed that Mt. Apo will have a sunrise two minutes earlier than Pusan Pt., as we had previously computed. I countered that our choice of Pusan Pt. was based on the presence of communities that are easily accessible to it. I further clarified that we had computed the sunrise times at Mt. Apo and the southeasternmost territorial boundary of the Philippines, near Dompoells Island which is on the Celebes Sea, that will have a sunrise a minute earlier than Mt. Apo. I also explained that sunrise is an event reckoned at sea level and is more appreciated by most of the people (Read http://www.travelsmart.net/ article/101181/.

On December 29, 1999, Carlito came to my office to interview me for the last time before I left for Pusan Pt. I mentioned to him rather casually that I will be coming with my wife there to welcome the new year together, which we have been doing since we married in 1969. I said, "I do not want to be separated with my wife on the first day of the year, so that we will be together the rest of the year". He wrote an article entitled "Astronomer mixes romance, science" that made it to the lower front page of the PDI on the first day of the new millennium! I had the article clipped, framed and hung on our living room wall for posterity.

Connie and I landed at the Davao City airport in the morning of December 30, 1999. We went to the DOST Regional Office and coordinated with my female friend who was the acting regional director, to borrow a vehicle and its driver for our trip to Pusan Pt. I promised to take care of the fuel expenses and the per diem of the driver. She gladly allowed us to use a DOST Pajero with *Jack Jacinto*, a 50ish white bearded man, as its driver. Jack and I agreed to travel the next day.

Nothing untoward happened when we made our first stop after three hours of travel in Mati City to have lunch and coffee afterwards. Midway into our second leg, we nearly had a fatal accident when our vehicle ran over a muddy patch while going up a slight incline and skidded toward a 5-meter deep rice field on the right. We could have been injured, perhaps seriously, if the vehicle plunged down the gully were it not for a single coconut tree lying on the roadside! The vehicle sustained a large dent on its front right fender and a flat tire. A group of men who were walking to the next town about a kilometer away saw what happened to us and helped pull the vehicle back toward the road. When we returned from Pusan Pt. the following morning, I noted that it was the ONLY coconut tree lying by the roadside for several kilometers in the vicinity. I then realized that God saved us from possible harm by putting that coconut tree at the exact place where our vehicle would have plunged! Praise God for His love!

We arrived in Santiago, the municipality in Caraga that included Pusan Point, at about 1 p.m. on December 31. We were met there by Gov. Lopez, who accompanied us to Pusan Pt., where we met the AsRDS staff who had left two days before. Elmor, Dario and Ruben have set up the telescope they brought and pitched a tent near the beach wall, facing the Pacific Ocean. The tent could only accommodate three adult persons, so I planned on sleeping in the Pajero with Jack and Connie. We had dinner in a make-shift carinderia in the area and tried to get some dozes afterwards. When it was about 11:30 p.m., I roused Connie and Jack to invite them to come with me to attend a pre-midnight mass in the area and to the tent of my colleagues to celebrate the new year eve. I brought with me a bottle of red wine and packs of peanuts to share for the occasion. Connie and Jack stayed outside the tent sitting on monobloc chairs provided by Gov. Lopez's staff.

We woke up quite early on new year day of 2000, excited to watch the first sunrise of the new millennium, which was computed to be at 5:45 a.m. on January 1. After having a simple breakfast at the carinderia, Connie and I joined my colleagues at the beach to wait for the time of sunrise. I thought that we were all praying that the cloudiness a storm in the Pacific was causing would dissipate and allow the sun to appear at that time. Our prayers were not heard. A local reporter and contributor to the PDI wrote a nostalgic piece about our personal encounter on the beach on new year eve. She seemed to have sensed in her report my pessimism about the situation.

Once again, as in the Total Solar Eclipse of 1995 in Tawi-Tawi, we went home humbled, but not humiliated.

Career Executive Service Eligibility (CESE)

I have been interested in the training for management being offered by the Development Academy of the Philippines (DAP) since my early years in the Weather Bureau. I have taken at least two qualifying exams prior to 1989 without success. I received an invitation from the Career Executive Service Board (CESB) to take the qualifying examinations for its Executive Leadership Program (ELP), when I became a division chief on May 4, 1989. I took the examination, which was on various topics from abstract reasoning, mathematics and reading comprehension and skill, among others. The test started in the morning and lasted for four hours, with one break of 10 minutes for snacks. The examination fee of Php 9,000 was authorized officially. I did not pass this exam. I applied again for the next one, this time I shouldered the exam fee and passed the test. The written exams turned out to be the easier one. The next phase of the ELP, the assessment center, was the more difficult part of obtaining a CESE. We were informed that more than 50% of those who were assessed do not pass this phase. I was fortunate to belong to the minority this time.

I, then, took three consecutive training seminars related to the ELP. The Salamin ng Paglilingkod, Session XLVII, was held at the Evercrest Golf Club & Resort in Batulao, Nasugbu, Batangas, on 28 February - 08 March 2001. Next came Diwa ng Paglilingkod, Session XXXIII, at Henry's Country Barns Resort and Hotel, Brgy. Masile, Calamba, Laguna, on 30 April - 06 May 2002. Finally, the Gabay ng Paglilingkod, Session XXIX, was held at the CESB Office, No. 3, Marcelino St., Holy Spirit Drive, Diliman, Quezon City, on 13 - 19 August 2002. I was conferred the CESE through CES Board Resolution No. 282 on 16 August 2000.

I am so proud and humbled to state that I have faithfully adhered to the pledges (See my CES PLEDGE at right) I made prior to the award of my CESE by the Career Executive Service Board. The words in bold and all capital letters are prescribed, while those in normal letters are our own selections from a number of choices.

The Philippines' Largest Telescope

I attended the UN/ESA Workshop on Basic Space Sciences: From Small Telescopes to Space Missions in Colombo, Sri Lanka on 11-13 January 1996. The workshop coincided with the inauguration of the 45-cm Cassegrain telescope, which was donated by the Japanese government to Sri Lanka, at the Arthur C. Clarke Center. I was keenly interested on how the donation took place and I inquired at the Center about it. I was fortunate to learn that three Japanese astronomers were in attendance during the inauguration and I approached them to ask how to avail of a similar donation from Japan. Upon returning to the

Philippines, as suggested by the three Japanese astronomers, I wrote a letter to the Japan International Cooperation Agency (JICA) applying for a donation of a similar telescope. JICA subsequently sent to PAGASA a form for the donation.

After almost two years later, two of the three Japanese astronomers, who I met in Sri Lanka in 1996, paid us a visit to look at the prospective site of the 45-cm telescope at the PAGASA Astronomical Observatory in UP Diliman. After several months, JICA informed us through a letter that our application for a donation of a telescope had been approved. From then on, I delegated to Dr. *Cynthia P. Celebre*, chief of the AsRDS, AGSSB, the process of the donation while I supervised the renovation of the telescope dome to accommodate the bigger instrument. Engr. *Dario de la Cruz*, chief of the Astronomy Observation Unit, took charge of the renovation. I have to state here that getting the telescope from the Bureau of Customs in Port Area, Manila was, to say the least, quite disgusting. We were not spared from the bureaucratic red tape by the Customs personnel, including their messengers and security guards who expected "grease" money to facilitate the flow of our importation papers. I had to shell out my one-month representation and transportation allowance to finally get the telescope out of the Customs area! It seems that they do not respect another government agency in their nefarious activities. It is quite regretful that my hope, President Rodrigo R. Duterte, for reforms in the BOC has also failed to change this notorious agency from corruption.

The new telescope was inaugurated in January 2000 and became the largest operational telescope in the Philippines. A similar smaller (40-cm) telescope is installed at observatory of the National Institute of Science and Mathematics Education (NISMED) of UP Diliman. Prof. *Edmund Rosales* was the astronomer in charge of the observatory.

 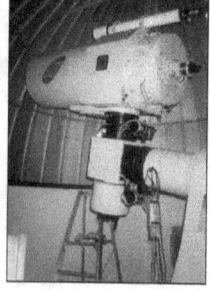

Figure 3: The 45-cm telescope installed at the PAGASA Astronomical Observatory.

PAGASA Astronomical Observatory (left) and 45-cm telescope (right).

Overhaul in PAGASA

PAGASA transferred to its permanent office building at the Science Garden on Agham Road, QC in March 2003. The building has four floors with the first floor occupied by the PAGASA Library, the Fellowship Unit, the office of the chief of the AsRDS, the Public Information and International Affairs Staff (PIIAS), and the CAB. The second floor houses the AGSSB chief's office, the Training Section, two training rooms, the Amihan conference room, and an extension office of the PIIAS. On the third floor are found the AD chief's office, the Personnel Section, the Records Management Section, the Procurement Unit, the Cashier Unit, the Field Operations Center, the Engineering and Technical Services Division, and the Financial, Planning and Management Division. The fourth floor is occupied by the Office of the Administrator and the offices of the three Deputy Administrators.

The other technical divisions of the PAGASA are at the WFFC, directly opposite the central office building on Agham Road. The Natural Disaster Research and Development Branch (NDRDB), the Flood Forecasting Branch (FFB) with its Flood Forecasting Center and the Weather Branch (WB), together with its Weather Forecasting and Communications Sections, are found there.

The three top executives of PAGASA retired one after the other, beginning in 2004. Dr. Amadore retired on his 65th birthday on July 1, 2004. Then, Atty. *Lilian G. Angeles,* Deputy Director (DD) for Administration and Engineering Services (A&ES), and Mr. *Ciprianoc C. Ferraris,* Deputy Director for Research and Development (R&D), both retired in 2005. DOST Undersecretary, Dr. *Florentino O. Tesoro,* was designated as Officer-in-Charge (OIC) of the PAGASA while Dr. *Rolu P. Encarnacion* and I were assigned as officers-in-charge of the offices of the DDs for R&D and A&ES, respectively. Mr. *Ellaquim A. Adug* was later named the OIC for the office of the DD for Operations and Field Services (O & FS).

The search for the new director had begun as early as 2003. It was followed in early 2004 by the search for the deputy directors of the agency. I was just a few months above the age limit set by the search committee at 60 years for applicants to the position but I applied just the same. I took the aptitude examinations on November 4, 2004, together with Dr. *Prisco (Boy) D. Nilo,* who much earlier furnished me a copy of his letter to the Director stating that I "was the most qualified to be the next director of the agency", and he is recommending me as such. Dr. Nilo was then the president of the Philippine Weathermen Employees Association (PWEA) and the Philippine Meteorological Society (PMS).

Drs. *Nathaniel Servando, Rosa T. Perez,* Nilo and Encarnacion, and Mr. *Martin (Jun) F. Rellin, Jr.* likewise filed their applications for the position. Drs. Encarnacion and Perez both also had the CESE and were more senior like myself in the service. During the selection process, I received a copy of a letter from the Career Executive Service Board to DOST Secretary, Dr. *Estrella F. Alasbastro,* informing her that Drs. Encarnacion and Perez and I have CESE and have priority in the consideration for the deputy director positions.

Prior to these, I experienced some events that may have become contributory factors against me. A letter was circulated in the office, suggesting that I may have an illicit affair with a fellow lady employee. It was followed by an email, supposedly sent by me to a few colleagues, with a similar suggestion. There was also a rumor that suggested that I am involved in corruption because I recommended the issuance of a certificate of completion of the MOTC to a weather observer who has been working for 20 years in Baler, Quezon. The observer did not complete the practical phase of the course earlier, due to reasons beyond his control. Lastly, someone furnished to Dr. Tesoro a copy of my "Hold" note (until funds are available) to the applications of some of my AGSSB officemates for monetization of their leave credits, on which my planning officer volunteered to write "This was disapproved by BMS" when the applications were forwarded to the Finance and Management Division for processing. She did not mention that the "hold" order was due to the information given by my finance liaison officer that there were insufficient funds for the purpose. That note, with the side comment of my planning officer, was shoved to me by Dr. Tesoro when he called me to his office, in response to a letter I wrote requesting the use of a vehicle for a personal trip to Nueva Vizcaya on a weekend. I assured him that I will take care of the fuel and per diem of the driver. He turned down my request. This incident further fueled my suspicion that there is a secret cabal in the office that wants to undermine the integrity of honest people in the PAGASA. I began to think of this when the DOST transferred about 100 of its rank-and-file personnel to the agency like people abandoning a sinking ship when public perception that the Department is a non-performing asset of the government and might be due for abolition.

I was disappointed to learn in 2005 that Drs. Nilo and Servando and Mr. Rellin were appointed as DDs for O & FS, R & D and A & ES, respectively. By then, Dr. *Graciano P. Yumul, Jr.*, who was another DOST Undersecretary, was appointed as OIC of the PAGASA. He was noted for his aggressive leadership and for soliciting many research grants abroad. His entry into the agency resulted in us working much harder than before due to the many projects he brought to the office. I thought then that he was a slave driver. He introduced us to the concept of branding to promote our products and services. He originated the PAGASA motto of "*Watching the skies ... serving the country*", which became our rallying point to do better in the performance of our duties and services. In support of this, I conceptualized the motto of the research and training and astronomy arm of the agency, the AGSSB, as "*Enhancing capabilities of sky and space watchers*".

Dr. Nilo rose to national prominence when he had controversy with Mr. *Gabriel (Gabby) S. Monroy*, who was his senior at the WFFC. Both of them were interviewed by separate tv reporters and each gave opposing directions for the center of a tropical cyclone that was then affecting the Greater Manila Area. Dr. Nilo gained employment in the PAGASA as a meteorologist while Mr. Monroy spent some time as a weather observer in a field station. Meteorologists are trained as weather forecasters, and basic knowledge of weather observation is taught only in passing. Weather observers are taught basic meteorology and methods of weather observation. The most important law related to determining the position of a tropical cyclone relative to the observer's position is the Buys-Ballot's Law, which states that when an observer's back is against the wind, the lower barometric pressure is to the observer's left in the northern hemisphere and to the observer's right in the southern hemisphere owing to rotation of the earth. They were caught on camera at the roof top of the WFFC building, pointing to the west and east when asked where the general direction of the tropical cyclone is. Mr. Monroy explained that when the observer's position is obstructed by tall structures like buildings and trees, the observer looks at the low cloud direction to determine the cyclone location. The WFFC is obstructed by the BIR building and the National Telecommunications Commission building on both sides. By the way, in meteorology, wind direction is the direction from which the wind is blowing. It is a logical definition because an observer would not know where the wind is going after it passes the observer's position.

Meanwhile, I researched the appointment of officials to the third level or executive positions and found that the Office of the President (OP) is the appointing authority for such positions. However, responsibility was assigned to the department secretary, who shall ensure that the recommended appointees possessed the necessary qualifications, including the appropriate eligibility. I wrote a letter, co-signed by Drs. Encarnacion and Perez and addressed to the OP, protesting the appointments of Drs. Nilo and Servando, and Mr. Rellin, Jr. We cited our CESE and longer services as our main justifications. The three appointed DDs have 27, 24 and 29 years of service, respectively. Drs. Encarnacion, Perez and I have 42, 39 and 41 years of service, respectively. We copied our letter to the DOST Secretary, the Career Executive Service Board and the Civil Service Commission for their information. The OP merely referred it to the DOST Secretary, who "prepared" a reply that Drs. Nilo and Servando had more accomplishments and completed more training, thus are most qualified for their positions. Dr. Yumul called me and Drs. Encarnacion and Perez to a meeting with Drs. Nilo and Servando at the Amihan Conference Room of PAGASA to settle (withdraw?) our protest. We reiterated our position while Dr. Nilo stood by Secretary Alabastro's letter. I questioned the Secretary's statement in her letter that Dr. Nilo had attended more training courses, because he was assigned at the Weather Branch for the whole duration of his 27-year service. All three of us were in the research and training divisions. I must mention that personnel at the WB were on call 24/7, especially when there is a tropical cyclone in the PAR. The meeting ended by Dr. Yumul advising us to see Dr. Alabastro and air our complaint to her. Following that, Dr. Perez came to see me in my office and said that she is no longer interested to pursue our protest because it had affected her emotionally. I, too, felt that we are being subjected to a whitewash and did not act as advised by Dr. Yumul.

The three deputy directors were appointed in acting capacity because they did not have the required CESE to be full-fledged directors.

Re-naming Philippine Tropical Cyclones

There have been several clamors prior to 1999 to change the naming system of tropical cyclones in the Philippines, due to various rights movements including the women's liberation movement. The current practice in the country was using Filipina names that are usually ending in "ing". Naming tropical cyclones and hurricanes after women began in 1953 when reconnaissance pilots gave the name of their wife and/or mother to identify the storm, especially when two or more were present to avoid confusion. The practice ended in 1978 in the U.S., which used male and female names in its lists for the Atlantic Ocean and Gulf of Mexico.

In response to a letter to the editor of the PDI on the issue of naming tropical cyclones in the Philippines, I proposed to the Executive Staff (ES) of the PAGASA to review and consider revision of its current list of tropical cyclone names, to include local female and male names and other terms native to the country. I reminded the ES of the importance of giving local names of tropical cyclones once it entered the Philippine Area of Responsibility (PAR). That is, a local name signifies that the cyclone has entered the PAR and is within three to five days of affecting of any part of the country. The public is, therefore, advised to monitor its development and to regularly watch for further information from PAGASA and advisories from the disaster risk reduction and management councils.

I suggested to conduct a contest to solicit possible names. The names that will get the highest number of proposals will constitute the new list. I drafted the rules of the contest, prizes for winners, and the members of the contest committee, which were approved by the ES. I was designated as chairman of the committee. The most salient rules I still remember are: 1) not more than 5 names with 15 letters; 2) local names of people (female or male), animal, flower, Filipino term; and 3) no offensive or unsavory meaning. One thousand pesos will be awarded for each name that will be selected for inclusion in the main list, which will be comprised of 100 names, and five hundred pesos for the 50 names in the auxiliary list. Both lists are divided into four groups, which will alternately be used every year, starting in 2001. An average of 20 tropical cyclones enter the PAR each year. When the actual number exceeds 25 in a particular year, the names in the auxiliary list are used.

Hundreds of letters were received at the PAGASA, submitting the proposed names for tropical cyclones. After much deliberation, the contest committee came up with the results (See table below).

Main List

Yr	2001, 2005, 2009, 2013	2002, 2006, 2010, 2014	2003, 2007, 2011, 2015	2004, 2008, 2012, 2016
1	AURING	AGATON	AMANG	AMBO
2	BAROK	BASYANG	BATIBOT	BIDAY
3	CRISING	CALOY	CHEDENG	COSME
4	DARNA	DAGUL	DODONG	DUGONG
5	EMONG	ESPADA	EGAY	ENTENG
6	FERIA	FLORITA	FALCON	FLOR
7	GORIO	GLORIA	GILAS	GILING
8	HUANING	HAMBALOS	HARUROT	HATAW
9	ISANG	INDAY	INENG	INGGO
10	JOLINA	JUAN	JUANING	JULIAN
11	KIKO	KAKA	KABAYAN	KENKOY
12	LABUYO	LAGALAG	LAKAY	LAWIN
13	MARING	MILENYO	MANANG	MANOY
14	NANANG	NENENG	NIñA	NONOY
15	**ONDOY**	OMPONG	ONYOK	OSANG
16	PABLING	PAENG	POGI	PANDOY
17	QUEDAN	QUADRO	QUIEL	QUINTA
18	ROLETA	RAPIDO	ROSKAS	RIGODON
19	SIBAK	SIBASIB	SIKAT	SIGLA
20	TALAHIB	TAGBANWA	TISOY	TOTOY
21	UBBENG	USMAN	URSULA	USA
22	VINTA	VENUS	VIRING	VIAJERO
23	WILMA	WISIK	WANG-WANG	WASIWAS
24	YANING	YAYAY	YOYOY	YOYONG
25	ZUMA	ZENY	ZIGZAG	ZOSIMO

Auxiliary List

26	ALAMID	AGILA	ABE	ALAKDAN
27	BRUNO	BAGWIS	BERTO	BALDO
28	CONCHING	CIRIACO	CHARING	CARAYAN
29	DOLOR	DIEGO	DANGGIT	DAGUNDONG
30	EKIS	ELENA	ESTOY	ESTONG
31	FUERZA	FORTE	FUEGO	FUERTE
32	GIMBAL	GUNDING	GENING	GARDO
33	HAMPAS	HUNYANGO	HANTIK	HARABAS
34	ISKO	ITOY	IROG	IKOT
35	JUEGO	JESSA	JOKER	JULIO

After we released the result of the contest in 2000, Sen. *Joker Arroyo*, who was running for re-election in the next elections, called to ask if his name was truly included in the list.   I answered his query positively and added that it may be advantageous to him in the coming election. He made a soft chuckle and thanked me.

The table was revised on May 9, 2005, when the PAGASA Officer-in-Charge noted that his nickname was in the list. He changed it to "Frank". He had other names changed, too, including "Yolanda" for "Yaning", apparently to make it appear that he was not only concerned about himself.

Quality Management System in the PAGASA

The World Meteorological Organization, in its last congress prior to 2000, adopted a resolution to institute a quality management system (QMS) among the world meteorological services. In compliance with the resolution, the PAGASA Executive Staff (ES) decided to develop its QMS and instructed me to search for a consultancy agency for the purpose. I found one from Makati City and contacted it. After agreeing to its terms, the consultancy agency sent a female agent, coincidentally with the same surname as mine, to discuss the procedures in establishing a QMS based on the International Organization for Standardization (ISO).   She patiently led me through the steps of writing the various standardized procedures to ensure high quality products as outputs of an organization. I was designated as the Quality Management Representative (QMR). I single-handedly wrote and completed the six manuals and the PAGASA Mission poster, which were approved by the PAGASA ES in October 2002. The manuals of the ISO 9001:2000 were titled: 1) Quality Procedure Control of Documents, 2) Quality Procedure Control of Records, 3) Quality Procedure Internal Quality Audit, 4) Quality Procedure Control of Non-conformity, 5) Quality Procedure Corrective and Preventive Act, and 6) Quality Manual.

I tried to obtain a certification from the ISO after this but lack of funds and material time hampered my desire. I was happy to learn that the Weather Division had obtained an ISO certification in 2015, using the six manuals I wrote. I earnestly hope that all the other divisions of PAGASA will follow suit as soon as possible, especially with the enactment of the Philippine Space Agency law on August 8, 2019.

Tatang, 77

A sad day for me came when news reached me about the death of my revered predecessor, mentor and benefactor, Mr. Catalino (Tatang)

P. Arafiles, on January 13, 2003 at age 77. At the time of his death, he was living in his hometown in Vigan, Ilocos Sur. He moved there from the Science Garden, QC, when his health began to deteriorate. His family arranged his funeral services to be at the Manila Memorial Park on Sucat Road, Parañaque City. I gave the only eulogy on the first vigil day at the memorial chapel. I wrote the eulogy hurriedly when my kumpareng Claro told me in the morning that I will deliver one when we go to the chapel in the afternoon. The whole Arafiles family from the US was present. They appreciated how I extolled their family head such that his son, Capt. *Virgilio P. Arafiles,* who was with the US Air Force, requested for a copy of my eulogy. I promised him that I will send the copy by email when I get back to my office to have it typewritten. He eventually had the eulogy published in the MIT Class 1950 Notes (alumweb.mit.edu/classes/ 1950.old/Jan_Feb_2005 notes.shtm). I would like to share my closing statement below because the last two sentences reflected my own attitudes toward work and life.

"Today, I pay homage to a great man who chose to serve his country than be with his loved ones, not because he had passed away but because he still lives in my heart. His work ethics are still my guiding principles. His integrity is my ideal and the honor that he has attained among scientists and his co-workers is my aspiration."

WMO RMTC in the Philippines

The Philippines began hosting the WMO Regional Meteorological Training Center (RMTC) for Regional Association (RA) V (South-West Pacific region) in 1969 with the implementation of a 5-year "WMO Training and Research Project, Manila" in the PAGASA. With the Department of Meteorology (DMO) in the University of the Philippines as its University Component, the RMTC aimed to meet the training needs of the country's meteorological personnel and to carry out research in various fields of meteorology. The PAGASA provided technical in-service training in various levels while the DMO offered a post-graduate course leading to a Master of Science degree and Doctoral degree in Meteorology. The RMTC for RA V had trained more than 2,000 Class III and Class IV meteorological technicians and around 1,000 Class I and Class II meteorological personnel since its establishment. Five of its graduates from DMO, who obtained a PhD degree, became the permanent representative of their respective countries with WMO. The RMTC produced more than 30 PhD and 100 MSc graduates, many of whom were employed in the PAGASA. I personally thought then as of today, that this may have contributed to the general improvement of the services of the agency since the inception of the RMTC in the Philippines.

In 2006, the WMO Executive Council decided that the term Regional Meteorological Training Centre (RMTC) should be changed to Regional Training Centre (RTC) to allow for specialization in areas other than meteorology, like hydrology and disaster preparedness. There are now 26 WMO Regional Training Centres composed of 38 components. The evolution of RTCs has resulted in a diverse portfolio of centres providing education and training through the use of residence classes, distance-learning and blended learning (https://public.wmo.int/en/resources/bulletin/celebrating-fifty-years-of-wmo-regional-meteorological-training-centres).

I was the WMO RMTC/RTC co-ordinator from 1982 until I retired in 2008.

Permanent Representative of the Philippines with WMO

I gained a distinct status in 2003 when DOST Secretary *Estrella F. Alabastro* instructed me to prepare a letter to the Department of Foreign Affairs, nominating me as the country's permanent representative (PR) with the WMO. I called the Secretary to inform her that it is necessary to name a representative due to the retirement of PAGASA Director Leoncio A. Amadore, who used to be the PR and the official link of the Philippines to the WMO. I suggested Undersecretary Florentino O. Tesoro to take the place but Sec. Alabastro told me to submit my name. I did as she instructed and WMO took note of it and sent me the WMO publication on the guidelines for PRs. Thus, I became the PR of the Philippines with WMO on July 16, 2003. My designation as PR was terminated on April 26, 2007 when a new PAGASA acting director was appointed. Mr. Martin (Jun) F. Rellin, Jr. was appointed as acting director despite being the least qualified among the three DDs, in terms of educational attainment and operational weather work experience, besides the CESE. In a speech during a flag ceremony at the PAGASA Central Office, a Congressman from Cebu added salt to injury when he said that before the appointment of Mr. Rellin, Jr., he conducted a poll in the agency that resulted in most of the personnel chose him to be the next director. Mr. Rellin, Jr. had a Master of Public Administration degree. He and his family left abruptly for Canada in November of that year.

Likewise, I became the second Filipino to be a member of the WMO Executive Council (EC) Panel of Experts on Education and Training in May 2003. Dr. Roman L. Kintanar, who was much respected and admired by the international as well as national meteorological community, was the first Filipino member of the Panel after he was elected President of WMO in 1979 and served as such up to 1987. The WMO President was the ex-officio Chairman of the Panel but he remained its chairman until 1995.

The first meeting of the EC Panel of Experts on Education and Training I attended was its 21st session in Antalya, Turkey, from May 3 to 7, 2004. Prior to this, I met a terrible accident on April 28, 2004 when I slipped in going down the spiral stairs at the PAGASA central office. It was then past 5 p.m. and the elevator was shut off when I took the stairs, together with Engr. *Catalino* (*Nonoy*) *L. Davis*, to join some of our friends in a joint outside the Science Garden. I heard a slight cracking sound from my right knee when I stepped down the stairs and I almost fell were it not for the railing that I held on. Nonoy asked if I was hurt but I said I was not and I told him that I would not be able to join him and our friends. I was able to drive home despite the slight pain on my knee. Upon arriving home, I asked Connie to accompany me to a doctor at the St. Jude Hospital and Medical Center on Dimasalang Road, Sampaloc, Manila. I informed the doctor that I was about to travel abroad in three days and I need something to enable me to do it without great pain. He injected into my knee something that relieved the pain on my knee. I left for Turkey on May 1 with a walking cane to protect my knee from further injury. I endured this condition until I was able to contact by email Dr. *Ruben P. Arafiles*, who treated free of charge my mother's foot earlier. Dr. Arafiles is the son of my predecessor, Mr. Catalino P. Arafiles. He was in the U.S. when I established contact with him. I informed him of my medical condition and he instructed me to send him my full lower body x-rays. He said that my right knee joint was misaligned by a few millimeters and he had to re-align the knee. After several months, he operated on my knee at the Medical Center Parañaque on Sept. 2, 2005. He placed a stainless metal brace in my right knee that is still in place up to today. He said that it has to be replaced after ten years, which may be the reason I feel occasional slight pain there. Dr. Arafiles and his wife who was an anesthesiologist gave me a 50% discount on their professional fees, for which I was grateful.

Maneuvering for the top position in PAGASA

I became an acquaintance of Jun Rellin when I met him as a weather observer in Cagayan de Oro City in 1977. He was a good looking and friendly person. When I learned that he was taking an engineering course, I advised him to complete it and apply for the Meteorologist Training Course at the Central Office. The next time I met him was at the TMRDO in 1999, when Connie and I joined the local chapter of the "Couples for Christ" (CFC). He was an active member of the CFC, who was impressive in his delivery of oral supplications. For this reason, I recommended him to Dr. Amadore to be his executive assistant vice Mr. *Eugenio Aquino*, who was a sluggish and an apathetic person.

I think that Jun's experience in the Office of the Director may have highly influenced him to aspire for that Office. Before his ascent as Deputy Director, he was rumored to be a relative of the Kintanars because he was from Argao, Cebu, the hometown of the political clan. When he was a DD, he called me to his office one day to show me a bunch of photocopied handwritten notes. The notes were in connection with the rumors that I recommended the issuance of a certificate of training completion to the weather observer who has been stationed in Baler, Quezon for 20 years. In fact, he was the Chief Meteorological Officer (CMO) when he wrote to request the certificate. The person who furnished Jun the photocopied notes intentionally did not include the intervening notes leading to my recommendation. The letter of request for certification was first indorsed by Mr. Gabby Monroy, who was the OIC of the FOC, and forwarded with favorable recommendation by Atty. Lilian G. Angeles, the DD (A & FS) to the Director, who in turn forwarded it to the AGSSB for final recommendation. I sent the letter to the Training Section for evaluation and recommendation. The Chief of the Training Section, Ms. *Efigenia (Jean) C. Galang*, wrote a memorandum stating the circumstances behind the applicant's non-completion of the required OJT for the MOTC and said that she cannot issue a certificate of completion. In response to the memorandum, Dr. Amadore wrote that he can sign a certificate, without any further comment. Based on the applicant's long experience as a weather observer, the fact that he was already the CMO in Baler, and considering Dr. Amadore's veiled threat to sign despite our opposition, I instructed Jean to prepare the requested certificate, which we signed together with Dr. Amadore.

Fortunately, the AGSSB kept photocopies of all memos, letters and other documents in its files. I copied the intervening documents which were deliberately omitted by the person who sent the incriminating papers to Jun. I sent all the related documents to Jun. I did not hear or learn anything from him after that, not even an apology from someone who is a fellow CFC member.

Years later, when Connie and I immigrated to Canada in 2011, Jun and his wife, Merly, did not come to the reunion of ex-PAGASA employees in the Greater Toronto Area, to welcome Connie and me. We had a brief chat when we met at the residence of *Gilda C. Borja*, one of our former office colleagues, in Scarborough, GTA. He came from work and said he was in a hurry to get home. He also did not talk with me when I called him to invite him to my 71th birthday luncheon on May 17, 2014. He said he will call me later as he was busy then at his work. He did not return the call. I thought that he was really avoiding me. I have forgiven, but not forgotten, him and Boy Nilo for their transgression against me.

Establishment of 1st Astronomy Course in the Philippines

Sometime in 2003, a man walked into my office lugging several volumes of books. He introduced himself as Dr. *Jesus Rodrigo F. Torres,* vice president (VP) for administration and academic affairs of the Rizal Technological University (RTU) in Mandaluyong City. He said he wanted to donate his books, entitled Urban Astronomy, to the PAGASA Library. I gladly accepted and took the six volumes he brought. We then had a chat over a cup of coffee. I learned that astronomy is just his pastime and does the observation of stars with his own telescope. He made observations in both the Pasig and Mandaluyong campuses of the RTU. He drew the position of the stars by hand, since he had no camera on his telescope. He wanted to show through his books that astronomical observation is possible even in urban areas which are highly light polluted. I wondered how he still got the energy to engage in his pastime, after his duties as VP and professor at the RTU.

In the course of our chat, I mentioned that I have been fascinated by astronomy myself since childhood but did not have the time and equipment to observe the skies at night. I wanted to study the science but found no school or university in the Philippines that offered the subject. Likewise, I told him that I wanted some of our AsRDS staff to get a formal course in astronomy to enhance their knowledge and skill in the science. I informed him that I wrote a letter to the National Institute of Physics in UP Diliman in 2002 to inquire if it can offer a course in astronomy there with the assistance of the PAGASA. UP replied that it does not have the budget and teachers to offer the course. Dr. Torres then suggested that I prepare a similar letter to the RTU president and he will favorably recommend it. I drafted a Memorandum of Agreement (MOA) between the RTU and the PAGASA, which was signed by Drs. *Jose Q. Macaballug* and *Prisco D. Nilo* in 2004. Dr. Torres drafted a course syllabus for a graduate course in astronomy and referred it to me for review and suggestions. The syllabus was adopted by the RTU board of regents and ushered in the first formal course in Astronomy in the Philippines, with the offering of a master's degree course in Astronomy or M.Sc. in Astronomy in RTU in 2005. I was appointed, together with Dr. *Cynthia P. Celebre,* Chief of the AsRDS of PAGASA, as a part-time Professor III, for the course. We encouraged three of the AsRDS personnel, namely; *Elmor A. Escosia, Jose Mendoza* and *Michael Bala* to apply for a PAGASA scholarship and enrol in the course. However, none of the PAGASA personnel completed the course, as of 2008. The first M.Sc. in Astronomy graduate was Dr. *Armando Lee* in 2010 (See photo below with Dr. Torres, then RTU President). He was followed by Mr. *Angelito Sing* in 2014.

In 2006, the RTU began offering a B.S. in Astronomy course with specialization in astrophysics, meteorological science or space science technology. As of 2018, a total of about 200 have graduated from the undergraduate course.

RP's first M.Sc. in Astronomy graduate, Dr. Armando Lee (right) with Drs. J.F.R. Torres (center) and B.M. Soriano (left).

OIC, ODD (A & ES) and Corruption in the Government

I was designated as Officer-in-Charge (OIC) of the Office of the Deputy Director for Administration and Engineering Services (ODD – A & ES) on July 10, 2007, in view of the appointment of Jun Rellin as acting Director, PAGASA. As such, I exercised supervision over the Administrative Division, the Finance and Management Division and the Engineering and Technical Services Division of the agency. I had the authority to sign and approve vouchers and checks not exceeding Php 5M in amount and other documents like requisitions and issue vouchers, applications for leave, etc. of the whole agency. In addition, I was named as chairman of the Bids and Awards Committee (BAC), which processed and recommended for approval bids submitted to the agency for various supplies, materials and equipment. I learned here the inner workings in the procurement process in the government service. Whereas before, it took a longer time to request and receive items of lower costs, I learned that those with higher amounts were purchased much faster because of the usual 10% commission given by the supplier to the agents of the buyer. This reminded me of a remark by Dr. Amadore who complained that buying ball pens takes longer to do than the tires or batteries of motor vehicles. I then imagined how the procurement process would have taken to purchase a surveillance radar system that costs Php 60M and how much the agent's commission was. There were rumors that some high-ranking officials in the national government demanded more than 50% of the total cost of an item or project. Even in the PAGASA, I heard rumors of an influential official who would withhold the processing of dealers' payment vouchers until they cough up with a relatively thick envelope. This practice became the standard operational procedure in the government, known as SOP later, from the national level down to the barangay level.

In this connection, I was likewise reminded of our observations when we surveyed the damages wrought by the passage of tropical cyclones over land. We observed then that public buildings and structures, like schools and roads, were the most readily or easily damaged after a tropical cyclone passage. We saw school buildings among old residential houses in the Bicol region destroyed while the houses remained intact.

I experienced another case of possible corruption when a provincial commander in Camarines Sur did not bother to ask for a "yellow" receipt for the Php 10K he gave to us as financial support for a seminar on disaster preparedness in Baao in the 1990s. A last story is about a revelation of my co-participant in the Marriage Encounter, under the CFC movement. My friend told me that he used to be a contractor in Samar. He related that when the birthday of the provincial or regional commander of the Philippine Constabulary was approaching, he would make ghost deliveries of supplies and materials to a government "project". The project's proponent would give the proceeds of the ghost deliveries to the commander as its gift for protection against harassments or lawless elements in the province or region.

In the sunset of my life ...

Retirement @ 65

In November 2007, less than six months before my retirement from the government service, I received a subpoena from the Regional Trial Court in Lagawe, Ifugao province, to attend as a government witness, called Friend of the Court, in the murder trial of U.S. peace corps volunteer *Julia Campbell*. Ms. Campbell was slain by a native of Banaue,

Ifugao, who claimed mistaken identity as his alibi. I was summoned by the Court to provide information on the lighting condition at the time of the murder, which was late in the afternoon of the month of November. Before coming to Lagawe, I consulted our astronomy staff regarding the position of the Sun, its altitude and azimuth, at the time of the murder. I was asked by the prosecuting lawyer if a person can be identified at the distance where he was standing in the court room (about 20 meters) at the time of the murder, which I answered in the affirmative. There was no further question from either side and I was excused by the judge. I do not know until the present time if there was already a verdict in the case.

I reached compulsory retirement from the government service on Saturday, May 17, 2008 when I became 65 years old. I spent 43 years 7 months and 5 days of my life in the Weather Bureau/PAGASA. Actually, it was exactly 44 years if my MOCTC training period, which started on May 18, 1964, was included. My AGSSB officemates prepared a birthday celebration for me on my last day in office on Friday, May 16, 2008. With the leadership of Dr. Carina G. Lao, they contributed funds to prepare the foods and decorations at the Amihan Conference Room of the PAGASA. About a hundred of my colleagues, including Dr. Prisco D. Nilo, current PAGASA Director, and Dr. Leoncio A. Amadore who retired earlier; Mr. Juanito F. Lirios, my second boss and former DD; Mr. Ellaquim A. Adug, ex-WB chief; Mr. Gabriel S. Monroy, my friend from Hydromet. Division in 1965 and WB Forecasting Center, and Mr. Rolando G. Valenzuela, my close friend and TGIF "glassmate". I value very highly the appreciation speeches given to me by some of my colleagues during the celebration.

My retirement celebration with all the guests

... and with the AsRDS staff.

Retirement is a time to reminisce what took place and what one has achieved during one's career. It is the time to look back at the "Monuments I have built in the sky". I have narrated here the most important events and achievements in my life and career in the government with utmost honesty and candor, based on my own personal recollections, with the aid of my curriculum vitae and occasional reference to the search engines on the Internet. I must mention that I have more than exceeded my life's goals, i.e. to complete my studies, to have a family and a home, and to travel abroad. My first goal is to get a college degree that will enable me to work professionally. I did not only get a college degree but also two graduate degrees. My second goal is a normal aspiration of any person. I achieved this goal, with a car as a bonus, and more importantly, together with Connie, I have fulfilled my obligation as a parent to see our children through college. I wanted to see some of the most famous world landmarks. Connie and I went to Lourdes (the apparition site of our Lady), France in July 1995. On our return trip to the Philippines, we made a side trip to Paris for one day to see and go up the Eiffel Tower and to visit the Notre Dame church. We went to see but did not enter the Louvre for lack of time. We also went to Jerusalem to visit the most important places in Jesus's life, including Bethlehem where He was born, the Vatican's St. Peter's Basilica (the seat of Catholicism), the Sistine Chapel, the Colosseum, the Trevi fountain (famous for the song "Three coins in a fountain" by the Four Aces), Italy's Venice (the canal city of Italy), Turin (where the Holy Shroud is kept), the United States' Grand Canyon, Statue of Liberty, Empire State building, and the 9/11 ground zero site in New York City, and Las Vegas, Australia's Sydney Opera House and Harbour Bridge, England's Big Ben, Tower

Bridge and Buckingham Palace, and many more. I made a total of 45 trips abroad from 1972 to 2008, 80% (or 36 trips) of which were funded by my sponsors or hosts. Connie and I used our personal funds to cover the nine trips. Connie joined me in every trip beginning in 1994. Earlier, I saw the Holy Shroud of Turin when I attended the Training Course on the Management of Meteorological Training Centres at the International Labor Organization training centre in Turin, Italy on 4 - 22 July 1988.

I earnestly hope that when the time comes to ask the questions "Was I brave and strong and true" (i.e. did I face adversity without hesitation or fear, was I determined to overcome my adversities, and was I faithful to my goals?) and "Did I fill the world with love my whole life through?" (that is, was I kind, affectionate, or benevolent to my fellow human beings?), God will answer them positively.

MHS 1959 graduation golden anniversary: preparatory activities

The first high schoolmate who I met after 48 years was *Higinio* (*Higgy*) *B. Simpliciano*, who came to my office one day in 2007. Higgy was my 2nd year classmate and was the youngest of our batch at 15 when we graduated from the Manila High School (MHS) in March 1959. His mother was the librarian at the Mehan Gardens campus of MHS. I invited him to lunch to celebrate our unexpected reunion. The idea to search for my high school batchmates started when *Estrella* (*Star*) *G. Siao* contacted me in FaceBook (FB) later. At first, I did not remember her until she said that she was the former Estrella Gaerlan, who was also my classmate in the 2nd year at the Manuel A. Roxas High School in 1956 – 1957. I remembered her as a petite young teen with dark eyebrows, hidden by a black rimmed pair of eyeglasses. She was very friendly then (until now). Upon recognizing her as my batchmate, we agreed to meet for lunch at the Bacolod Chicken Inasal restaurant at the QC Memorial Circle. During lunch, she mentioned that she had been in touch with *Nila Nayo Santos*, my classmate in Section 1, and *Adelaida* (*Aida*) *Caluyo Bermas*. I asked Star to arrange a lunch date with the two ladies, together with Higgy, at the Chinatown Restaurant on Banawe Ave., QC as soon as possible. Nila informed us when we met that she knew that our class valedictorian, *Rosalio* (*Rhoti*) *P. Torres*, is now a doctor of medicine and is working at the Makati Medical Center (MMC). Subsequently, I searched for his name at the MMC website and found him to be a hematologist. I tried to call his clinic but no one answered my call. I left a message on the answering machine, giving my name, office phone number and email address. After a few weeks, Rhoti called me. I informed him that I have met some of our high school classmates and batchmates. He asked to arrange a lunch date with them at the Chinatown Restaurant. We had a happy reunion during lunch later and talked of our forthcoming graduation golden anniversary in 2009. I volunteered to search for other schoolmates to invite them to join our discussions for our 50th graduation anniversary.

Star and Bernie at Bacolod Chicken Inasal, QC Memorial Circle

We had another meeting at the Chinatown Restaurant for dinner with Rhoti, who came after his rounds of his patients in the Cardinal Santos Medical Center in San Juan City. We discussed during dinner the preparations we had to make for our graduation golden anniversary. It was at this moment when we decided, among others, to establish a collegiate scholarship program for the alumni of the Manila High School, knowing that students of our alma mater came mostly from poor families. Our group, composed of Rhoti, Star, Nila, Aida, Higgy and myself, became the MHS '59 scholarship board of trustees. We selected five, one for each decade after our graduation, MHS alumni after thorough analysis of their academic performance and personal background. Our first scholars were *Leonila (Oni) Herrera, Daisy (Daize) Calupas, Cristina (Cristy) Dapulag, Deric Arediano* and *Camille Casela*. *Bernie Alber, Joanne Baao* and *Laurice Baao* replaced two scholars who graduated from college later. We provided the scholars a modest monthly transportation allowance with pocket money and laptop computers, which were given by Rhoti. We had frequent meetings, hosted by Rhoti at his residence, to celebrate birthdays, Christmas, new year, and other special occasions, where we invited all our scholars to attend. MHS '59 is proud and extremely satisfied to have been a significant part of the personal development of nurse Oni, RCBC OJT Daize, OFW (Malaysia) teacher Cristy, architect Deric, teacher Camille, engineer Bernie and architect Joanne. Up to this day, we continue to select deserving MHS alumni for our scholarship program.

Aida, Nila, Star and Bernie at Chinatown Restaurant, Banawe Ave., QC. Taken on May 24, 2007 at 1:04 PM.

MHS '59 alumni with four of its scholars after receiving their laptop from Dr. Rhoti P. Torres with Dr. F. Caluyo, brother of Aida, as guest.

Daize Calupas receiving her laptop from Dr. Rhoti.

Higgy invited me to search for the names and addresses of the batch of 1959 at the Manila High School in Intramuros, Manila. We talked with the MHS principal, Mr. *Arnulfo Empleo*, to request permission to look at the school records. He gladly acceded to our request. I first met Mr. Empleo during the centennial celebration of MHS in 2006 at the Manila Hotel, where I came to be acquainted with two batchmates, *Eduardo (Ed) C. Maagma* and *Joselito (Joey) D. Justalero*. I learned that the original Manila High School started in 1906, with four campuses, one each at the northern, eastern, southern and western part of Manila. The new Manila High School was the one on the western part of Manila.

Fortunately, we found the roster of 1959 graduates, including their address. We borrowed, photocopied and properly returned the roster immediately. I patiently typed the roster, comprising of 636 names spread among 15 sections, into an electronic table, which I used to search using google.com for the names in the list. Eventually, I found 41 of my batchmates, who have postal addresses, telephone numbers and email addresses. Among those I found, in addition to the six in Metro Manila, are: *Rodolfo (Rudy) C. de la Cruz* and *Virgilio (Biyo) S. Lopeña*. Rudy was a businessman, who owns a funeral parlor in Cavite City. Biyo was a retired Makati City policeman, who is in charge of the security of the subdivision where his house is. I contacted most of my 41 MHS co-alumni to invite them to our golden anniversary reunion. I informed those who were in the U.S. of my plan to go to Canada in August 2008 to visit our daughter and her family and to the U.S. to visit my relatives later. They suggested to meet them in Las Vegas on October 18 to have a mini-reunion, to which I agreed.

MHS '59 scholars Bernie, Oni, Camille, Cristy, Deric, Joanne and Laurice

With their sponsors and Connie. Daize was absent.

I first contacted *Priscilla (Precy) Jimeno Mitchell* by email after Higgy, who had his parents living in California, returned from a visit there and told me that some of our high school batchmates are also living near his parents' home. He mentioned Precy's name who was my classmate in Section 1. I searched for her maiden name on google.com and found her to be in Pittsburgh, CA. It is quite convenient to look, through the internet, for people who have their personal information after having worked or had business in one's country. We exchanged messages since I found her and she gave me the general location of her former boyfriend, our high school salutatorian, *Consuelito* (*Lito*) *U. Legaspi*. Again, I found Lito's business address and contacted him. I provided both Precy and Lito our 4th year class picture and roster of 1959 MHS graduates by email. Precy likewise provided me the address of *Myrla B. Miguel Espineli*, also our 4th year classmate. Subsequently, I located *Gabriel M. Abad, Jr., Helen A. Arceo del Rosario, Susan Arceo Bondad, Hilario G. David, Walter B. Feir, Aurora* (*Auring*) *P. Tobias Gajilan, Ricardo* (Ric) *F. Garcia, Josefina* (*Josie*) *Reyes Garcia, Ernesto* (*Ernie*) *C. Liwanag, Angelita* (*Lita*) *C. Liwanag, Carmelita* (*Carmen*) *Losmagos Diffin, Jacinto* (*Jack*) *L. Marquez, Jr., Esperanza* (*Espie*) *Capacete Magno, Purita V. Manlapaz Ramos, Renato* (*Rene*) *M. Morales, Aida M. Apolinar Piring,* and *Lourdes M. Umali.* Hilario was a physician in Florida, who did not respond to my message on his answering machine. I tried to locate my 4th year classmates *Alfredo P. Carracas* and *Rodolfo R. Obiniana* but was unsuccessful. I found Walter's workplace but learned that he had left his work. I was saddened to find that *Jacinto L. Ledesma* had passed away a day before I called his telephone number and I learned that Rene Morales had also died when I found his address. It was much later when we located *Rosalina* (*Lina*) *Gan Uy, Teresita* (*Tess*) *Santos, Alicia* (*Lizzy*) *Osio Yumang, Mila V. Parilla, Ricardo O. Causapin,* and *Erlinda C. Custodio Tolentino.*

Precy surprised us when she made an unannounced visit to the Philippines from June 27, 2007. She contacted me via email to meet her in Glorietta 2 Mall on June 28. I informed the others to come at the meeting. Seven of my MHS batchmates came to meet our co-alumnus from the U.S. We all had lunch at the Glorietta Mall, then we proceeded to Biyo's house in Taguig City, Rizal for merienda, especially "balut".

Standing: Biyo, Rudy, Precy, Aida, Star, Bernie in Glorietta 2 Mall. Kneeling: Higgy & Joey

Rhoti in tie is added, in Biyo's house in Taguig City, Rizal.

Rudy invited us to his rest house in Amadeo, near Tagaytay City, Cavite to have another mini-reunion on June 30. Connie and I met Nila at the McDonald's along EDSA near the corner of Quezon Avenue, then drove to Tanauan, Batangas where Precy lived with her relatives. We met our other batchmates at the Tagaytay City rotunda to have snacks, before proceeding to Rudy's house where we had lunch.

Precy then invited us to meet at the Mall of Asia (MOA) in Pasay City on July 5. I found through Joey another batchmate, *Raul M. Jaque*, who joined Precy, Nila, Star, Aida and me with Connie, for the first time. Raul lived in a subdivision close to Joey's place in Parañaque City. We had lunch at Max's restaurant, where Precy had halo-halo for dessert, her favorite she said.

Star and I soon began preparations for our reunion by searching for possible venues online after Precy returned to the U.S. Aida and Higgy suggested two sites, which we found unsuitable for our purpose. We found that Kabayan Beach Resort in Ilaya, San Juan, Batangas is the best for our class celebration. Higgy, who is a free-lance artist, designed the batch logo and souvenir magazine cover. After having the logo agreed upon by us, we called the Manila group, Star ordered two sets of shirts (one yellow and another green colored) with collars and breast pocket with the logo printed on it. She ordered 50 pieces of each color. She also ordered two sets of caps with the same colors and logo. She had two tarpaulin banners made for decoration during the celebration. We all agreed that we would deposit Php 2,000 each as our initial contribution for the reunion. We informed the US group of our arrangements, including the contribution. They agreed and promised to cover the remaining costs for the celebration.

In the meantime, we practiced at least three times on different occasions a dance number at Star's residence in Proj. 2, QC. We also advised our US counterparts to prepare a dance presentation, too. Photo below shows Nila, Aida, Star and Lizzy after our dance rehearsal at Star's house, holding a tarpaulin banner prepared and printed by Higgy.

Nila, Connie & Precy on Talisay Road with Taal Lake at the background. Taal Lake has the world's smallest active volcano, which has another volcano within it

At Rudy's rest house in Amadeo, Cavite.

US Tour

Our daughter and her family have moved to Brampton, Ontario in 2007. Connie and I embarked to tour the U.S. starting on September 20, 2008. We took a Greyhound bus from Toronto to visit first my former planning officer and MTC classmate, *Arnulfo (Noly) Q. Bolante* and his wife, *Geronima (Ninin) Biason*, who was likewise my former officemate, in New Jersey. Noly and Ninin drove us to New York City the following day to join a cruise aboard the "Spirit of New York" on the Hudson River around the famous landmarks on the river, particularly the Statue of Liberty and the Ellis Island immigration building. After the river cruise, we went to the Empire State building and passed by Broadway Avenue, before going to the ground zero site of the World Trade Center (WTC) buildings, which were totally destroyed when two planes commandeered by terrorists crashed into the buildings on September 11, 2001. Construction of the new WTC buildings was going on.

Noly and Ninin drove us on the early evening of September 24, 2008 to Jersey City to visit my first cousin, *Catherine (Cathy) V. Evidente*. Our auntie Lulu and uncle Naning were living with her on a 3-storey building she owned. Cathy invited our Valdez cousins, who lived in New York City and New Jersey, to a welcome party in the evening of the following day. Jun and his wife Mila, who was also an MD, and Sandy and Sonia came, together with Cathy's two daughters, Cindy and Divina, and their boyfriends. We had a grand evening dining and singing with the videoke.

Cathy brought us, together with auntie Lulu and uncle Naning, to Borgata Hotel in Atlantic City the next day. She booked all of us in two separate rooms free of charge, which were part of the privilege she earned due to her frequent visits to the hotel casino. It was our first experience to go to a casino and Connie and I played the slots. We lost US$20 each and stopped. Auntie Connie, who lives with her husband, *Joseph (Joe) Byrne*, in Alexandria, Virginia, came to fetch us (Connie and me) in the morning of Sep. 26 to spend two days with her and uncle Joe.

On Sep. 27, auntie Connie dropped us at the Library of Congress in Washington DC, then picked us and drove us to the Lincoln Center. We had a few photos taken in each of these tourist spots. She said she stayed in a parking spot nearby and read a book while we were walking around. She gave us an hour or two for each before she came for us. We truly appreciated her effort in giving us a pleasant visit to the nation's capital and her home.

Auntie Connie brought us on Sep. 28 to visit Arlington National Cemetery, where famous American presidents, particularly President *John F. Kennedy*, and war veterans were buried. We watched the solemn ceremony of "Changing of the Guard" at the Tomb of the Unknown Soldiers. The Tomb has been guarded by a sentinel every single second of every day since 1937. Through hurricanes, blizzards and blazing hot summers, the perfectly-dressed guards keep watch on the tomb on 24-hour shifts. The guard takes 21 paces in one direction across the tomb, pauses for 21 seconds facing east then 21 seconds facing north, takes 21 steps back to the other side of the tomb, then repeats. Each time he turns, he snaps his heels together, making a clicking sound, and shifts his weapon so that it stays facing the visitors. The significance of 21 relates to the 21-gun salute, the military's highest honor (https://www.whiskiedwanderlust.com/changing-of-the-guards/). The ceremony, that takes place every hour at the top of the hour during winter (and every half hour from April to September), was so solemn that nobody was allowed to make any sound or noise during the changing of the guard. I read that the guards who are assigned for duty there are carefully selected and they undergo rigid training before being posted. We also went to take a look at the grave site of Senator *Robert F.* Kennedy, JFK's brother, and *Audie Murphy*, one of the most decorated American soldiers of World War II. I admired JFK for his significant role in American history. I particularly liked JFK, not only for his inaugural speech when he said

"Ask not what your country can do for you – ask what you can do for your country", but also for spearheading the world to space exploration. I will never forget the date he was assassinated in Dallas, Texas on November 22, 1963. This was also my motive in naming our third child Johann, whose birthday coincided with JFK's assassination.

Kennedy family grave site

Tomb of the Unknown Soldier

We next visited the Smithsonian Institution, a museum and research complex in Washington DC. We first went to the Museum of

Natural History where we saw and took pictures of the artifacts there, including the mammoth, dinosaur skeletons, and a sperm whale

Auntie Connie drove us to my high schoolmate, *Cristina Lopez Ulis*, who lives in Annapolis, Maryland on Sep. 29. Together with her husband and auntie Connie, we roamed around the US Naval Academy, which is famous for its excellent facilities and training of naval personnel.

I booked two online plane tickets for our next stop at Tampa, Florida, to spend two weeks (Sep. 30 – Oct. 14) at Sandy's home in Wesley Chapel, Fla. We almost moved to a hotel, after a week there due to awkwardness for staying too long in my cousin's abode. I told her that we would stay in her house for that long because I was meeting my high school classmates in Las Vegas on Oct. 18 for a reunion. She seemed generous when we first met in 1998 in New Jersey because she offered me a plane ticket to the Philippines. Anyway, during our stay in Wesley Chapel, Sandy introduced us to her next-house neighbor, *Pachot*, an elderly lady who is so good at piano playing. She enjoyed playing the musical instrument when we went to her house. Sandy also showed her piano playing talent there. By the way, Sandy and her siblings Jun and

Sylvia were likewise quite good in painting. Jun completed a Bachelor of Fine Arts course in UP Diliman when he was already past 60. Cathy, who also had a house just a few houses from Sandy's, followed us later in Florida. She drove all the way from New Jersey, with auntie Lulu as her lone passenger. All five of  us went to Orlando, Fla. to see the Disney theme parks there. However, we only walked around the Disney World and Legoland and had a few photos taken. Connie and I joined a day tour from Orlando to the Kennedy Space Center where we saw and took photos of the giant rockets (Photo above) and other relics of NASA's space program, including a re-entry space capsule and the rocket launching pads.

We left Florida by plane on Oct. 15 and arrived in Phoenix, Arizona in the afternoon to visit Dr. *Jacinto* (*Jack*) *L. Marquez, Jr.* and his family in Peoria, Az. Jack was my batchmate in the Manila High School.

He is now an anesthesiologist and his wife, Tessie, is a nurse. They brought us the following day to Scottsdale, a city 31 miles (50 kms) east southeast of Peoria. Scottsdale seemed to be an old Indian settlement with stores selling products made by native Americans. We went to a Filipino restaurant there and had lunch and halo-halo.

We drove the next day to Las Vegas, where we were to meet some of our batchmates for a mini-reunion beginning on Oct. 18. We passed by the Hoover Dam road, where a new bypass bridge was under construction, on the way to the west side of the Grand Canyon. We stopped there for a few hours, taking pictures and looking at the attractions there. We did not try the new Sky Walk, a short arch with transparent glass floor protruding out of the cliff, because it costs too much. The entrance fee is US$30, plus another US$30 if you bring a camera in to take photos.

Connie and Bernie at Eagle Pt

Jack and Tessie with the mighty Colorado River that carved the Grand Canyon.

We arrived in Las Vegas in the afternoon and checked in at the Circus Circus Hotel Casino, where some of our batchmates have previously billeted themselves, including *Emilio (Emil) Ermio*, the vocational class valedictorian of the 1962 MHS batch, who wanted to join our batch reunions. We met there *Aida M. Apolinar* and husband, Mr. *Piring*, *Priscilla (Precy) A. Jimeno-Mitchell*, *Aurora (Auring) P. Tobias* and hubby *Lito Gajilan*, *Myrla B. Miguel* and husband *Cecilio (Kuya)* Espinelli, and *Consuelito (Lito)* U. *Legaspi* and wife *Grace*.

I told them that we have tentatively set our 50th graduation anniversary celebration at the Kabayan Beach Resort, Ilaya, San Juan, Batangas, on March 16, 2009. We enjoyed each other's company after not having seen or heard from each other for the past 49 years! I told them how I found our other batchmates earlier in Pinas and they were so glad that I did, just in time for our golden graduation anniversary.

Connie and I spent our first full day in Las Vegas walking around the various casinos that lined the "Strip". We were so awed by the beauty of Hotel Bellagio, which had beautiful interior decorations and the only dancing fountain on its front. We had dinner in a restaurant on our first night in

LV. We went to the Mirage hotel to watch the Cirque du Soleil perform "The Beatles Love" on Oct. 19. On our third night, most of the ladies went to Lake Tahoe with their partners. Connie, Aida and I did not join them. We all went our separate ways after breakfast on Oct. 21, promising to see each other again in our graduation anniversary celebration in 2009. Before parting ways with Precy and Myrla, I informed them of our plan to visit my relatives and a friend in California and Seattle. I asked if we could pass by their respective homes in California for a few days in route to Seattle. They both readily and happily agreed.

We took a Greyhound bus on Oct. 21 from Las Vegas to Fontana, California, where my cousin *Vicki F. de la Cruz* and her family live. Vicki and her husband, *Federico (Boy) de la Cruz*, met us at the bus depot. I had previously arranged with Vicki's brother, *Napoleon (Nap) S. Ferraris* and younger sister, *Helen F. Serneo,* that we would spend some nights with them. Nap had a house on Adolphia Ct., San Diego, CA while Helen lived with her husband, *Federico (Freddie) Serneo,* in Chula Vista, CA. Nap was a retired US Navy (USN) Commander and was a civilian employee in the USN when we went to California.

We slept for three nights (Oct. 21-23) at Vicki's home, where we got acquainted with her eldest child, Jessica and her two daughters. Jessica's older daughter, Kayla, was about three years old while her sister, Brianna, was only three months old. Kayla was quite talkative and friendly to us. It was only much later that I came to know on Facebook the name of their father, who is *Malchiah Heads*. We hardly talked with the two sons of Vicki and Boy, who are *Jeffrey* and *Jeremy,* because they were out most of the time we were there. Vicki and Boy drove us to Los Angeles, CA to see the Griffith Observatory on Oct. 22. The road leading up to the Observatory is where 8-time world boxing champion, now Philippine senator, *Manny Pacquiao* did his road work to prepare for a fight in the U.S., usually at the MGM Grand Hotel Casino in Las Vegas. I was so impressed at the facilities of the Griffith Observatory, especially its Foucault (pronounced "Foo-koh") Pendulum and planetarium, which was much better although older than ours.

With Vicki and Boy and their granddaughters

At Griffith Observatory, LA

We next stayed in Nap's home (photo above) in San Diego on Oct. 25-26, 2008. Nap and his wife Grace tendered a welcome dinner for us, with the De la Cruz and Serneo families. Our cousin, *Rey S. Tito* and his wife *Grace*, who live in Chino Hills, came later. Rey is the youngest child of my auntie Juliet and uncle Seniong. He invited us to spend a day in his home before he and Grace left after the party. While drinking beer and having a karaoke session after dinner with Nap, Boy and Freddie, I remarked that Boy and Freddie have the same first name. They both laughed and said that Vicki's and Helen's youngest sister's husband is also named Federico, with the nickname Fred. We all laughed at realizing that the three sisters have married men with the same first name!

Nap brought us in the afternoon of Oct. 25 to a farm near San Diego to go persimmon picking. We enjoyed picking the fruit which cost us only US$5 a five-gallon barrel. We brought home two barrel-full of persimmon in bags and shared the fruits with our cousins and their families.

Our next stop was at Helen's home in Chula Vista, on Oct. 26. Freddie and Helen brought us first to Balboa Park, then to San Diego to see the USS Midway museum. Freddie was a retired U.S. Navy man. Before going home, we passed by the commissary to buy my black leatherette jacket, which I wore in San Francisco later.

Rey came on the morning of Oct. 27 to pick and bring us to his house in Chino Hills, CA, by far the largest of my cousins' houses in California. He and his wife Grace have three young sons, named Iphraem, Nathan, and Samuel. Rey and Grace both worked in the Kaiser Hospital as a medical coder and bills analyzer, and nurse, respectively. Nathan was an aspiring chef, who professionally prepared all the foods we shared for dinner when we arrived. We enjoyed all the dishes he prepared and I think that he will become a famous chef one day. Rey drove us on the same day to Torrance, CA to meet our auntie Nita and uncle Dr. *Jose (Joe) Soriano*, cousin of my father and Rey's mother. I have not heard or seen Auntie Nita for at least 50 years since I was in Del Pilar St., Singalong, Manila in 1957-1958. She has remarried and is now *Juanita Quellenberg*. She came with her son Modesto Jr. and daughter Jasmine. The last time I saw uncle Joe, who is a medical doctor, was when he was in Makati City at the Gilarmi Hotel in the 1970s. We had a happy reunion over lunch with two of my immediate ancestors in Torrance, CA.

On the morning of Oct. 28, another Rey came to fetch us to bring us to his home in Glendale, CA. This time, it was *Reynaldo (Rey) P. Munda*, my MTC classmate and "kumpare" (my son Chris' godfather). He checked us in Days Inn Hotel first and brought us to his home to have lunch with his wife Linda. We sang with his karaoke set while having coffee after lunch.

Rey brought us the following morning to Griffith Observatory, then to Hollywood to see the sights and the Universal Studios in the afternoon. Many of these activities were recorded by Rey with a still digital camera on his left hand and a video camera on his right hand. We walked on the famed (pun intended) "Walk of Fame" and passed by the Kodak Theater where many Oscar awarding ceremonies were held. We then proceeded to the Universal Studios, where we saw different movies sets, including those for Jaws, Water World, Backdraft, Jurassic World, and Fast and Furious, among others.

The next leg of our U.S. tour was to Pittsburg in northern California to stay in Precy Mitchell's house. Rey drove us to the Ontario (not Canada) international airport in the Greater Los Angeles Area for our flight to Pittsburg on Oct. 30. He said that I could have chosen the airport closer to Glendale but he did not mind the longer travel to Ontario. I tried to learn how to take a selfie with a digicam, as Rey did in Hollywood, while waiting for our departure. I was successful to get a photo of myself on my third try.

Precy picked us up at the Pittsburg airport and drove us to her home where her older sister, *Luzviminda Velasco*, was staying with her. She and her sister worked in a department store in Pittsburg as salespersons. Both of them are widows with a son each who are already adults and living separately from them.

Precy drove us to the house in Roseville, CA of Myrla and her husband Cecilio or Kuya, where we had lunch. We then picked up *Enrique* (*Ike*) *B. Cuaño* and his wife Blanca from their home in Vallejo, CA. Ike was also our MHS batchmate, who I met for the first time. Together with Precy, Myrla, Kuya, Ike, and Blanca, we later met Emil, who likewise lived in Vallejo, and visited her mother in a hospice home there on Oct. 31. We then proceeded to San Francisco (SF) to view the sights, most notably the Golden Gate bridge and the Fisherman's Wharf, where we had a taste of its famous clam chowder soup made more welcome due to a chilly, rainy afternoon.

After satisfying ourselves with the sights and foods in SF, we all drove to Napa Valley, CA, for a wine tasting tour and a visit to the petrified trees. We made only stops at two wine factories, where we had a glass quarter filled with their finest wine for free. No one bought a bottle, 'though.

We then drove to the world's largest petrified forest, where large trees that have been buried for millions of years have turned into stones were unearthed in the Napa Valley. Photo at below shows Connie beside The Queen, a 65-ft long, 8-ft in diameter coastal CA redwood, 2,000 yrs. when living, buried approx. 3,400,000 yrs. (Photo below).

Pose at the crookedest road in the world in SF

Connie and Bernie pose with the Golden Gate bridge in the background.

Connie and I slept at the Cuaño home in Vallejo for that night and Precy fetched us the next morning to bring us to the Pittsburg airport for our onward flight to Seattle, Washington to visit my maternal relatives there.

We were met at the Seattle-Tacoma airport by my cousin, *Elmer* (*Elm*) *L. Monillas*, who used his large GMC van to bring us to his parents' home in Seattle, where we stayed from Nov. 4 to 7. Elm had at least six different expensive cars besides the van in his garage, including a hummer, a Porsche sports car, a BMW plus two others. He used the van to bring us, together with his parents and elder brother, Edgar, to west Seattle, where its skyline is best viewed with the Space Needle. He also brought us to visit his eldest sibling, Malen, and her husband *Mike Fulk*, and to a night club where impersonators performed. I was impressed by the singer who closely looked and sang like the legendary Ray Charles!

With Auntie Tessie and Uncle Celso in west Seattle

Uncle Raul fetched us and drove us to visit his elder brother, Berting, who he has sponsored to a hospice home near Bothell, WA where he and his family live. He also brought us to the columbary, where my maternal grandmother's ashes urn was deposited. Nanay was two days short of 100 years and five months when she died on April 13, 2004.

Connie with my youngest cousin, Lora, uncles Raul and Berting.

Elmer and Edgar brought us to the Greyhound bus depot in Seattle on the morning of Nov. 8 for our trip to Vancouver, BC, Canada. A minor incident occurred at the queue for checking in when I left the line, leaving our luggage to attend to a personal matter. Upon my return, the person, a black man, behind me softly complained that I inserted myself in the line. I said I was standing there before he came, pointing at my luggage. He said, he did not see me when he arrived. Elmer and Edgar, who were standing behind, noticed our slight commotion and approached us, saying "What's the matter, man?" Being bigger and taller than the man and his companion, he did not say anything. I waved my cousins back and I stood behind the man.

Connie and I arrived at the Greyhound depot in Vancouver and called *Phanie Sanchez*, daughter of our ACF Homes neighbor, Jesus and Precy Sanchez. She invited us beforehand to visit her and her family in Coquitlam in the Vancouver area, on our return to Toronto. Her husband picked us up at the depot. Phanie introduced us to her father-in-law, who offered me a beer together with his son. Afterwards, we had dinner of the boiled in water Dungeness crab which Connie and I enjoyed so much because its meat was fresh and sweet tasting.

The final leg of our tour was to Surrey, Canada, to visit my cousin *Rosanna M. Belamala* and her family. She fetched us from Phanie's house on the morning of Nov. 9 and we went to her house, where we met her daughter and son. Her husband, who was a 16-wheeler cross country driver, arrived later and  we had lunch together. Rosanna and her husband drove us to Stanley Park in Vancouver (Photo at right), before going to the airport for our homeward flight to Toronto.

Thus, our 11-week tour of the U.S. ended on Dec. 8, 2008. The tour took us through the states of New Jersey, Virginia, Washington DC, Maryland and Florida on the east coast; Arizona and Nevada in southern U.S.A.; and California and Washington on the west coast.

MHS 1959 graduation golden anniversary: G-Day

Only 16 alumni, some with their spouse, were able to attend the reunion on March 14 to 16, 2009. Five came from the U.S. while one arrived from Finland and the rest from Manila. Precy Mitchell arrived first from the U.S. about a week before our grand reunion. She stayed in her hometown in Tanauan, Batangas. We all took a tourist bus in the morning, picking up others at pre-designated locations along the way to Ilaya, San Juan, Batangas on Saturday, March 14. We arrived near noon at our destination after negotiating the South Luzon Expressway, which was being renovated then.

We enjoyed tremendously the two and half days spent at the beach resort for our golden graduation anniversary reunion. We were like the teenagers that we were half a century ago at the Mehan Gardens campus of the Manila High School. We had lunch at the beach on our first day, then took some souvenir pictures and later enjoyed singing with a karaoke and dancing in an open beach pavilion. In the evening, we had a luau dinner at the beach, complete with torches and us wearing floral dresses and shirts, some with leis. We all went after dinner to the pavilion where we will have our program and presentations on its second floor. The presentations began after the welcome remarks were given by our class valedictorian, Dr. Rhoti P. Torres, and our other batchmates, including Precy, Myrla, Helen, Purita, Star, Nila and myself. The Manila group first performed with the dance we rehearsed at Star's house. The U.S. group with an all-women ensemble wearing long floral dresses with leis around their necks, was impressive with its hula dance to the medley of "Tiny Bubbles" and "My Little Grass Shack". We learned later that they rehearsed their dance number in Honolulu on the way to the Philippines from the U.S.

MHS '59 on Mar. 14, 2009

Mar. 15, 2009 in KabayanBeach Resort, San Juan, Batangas.

The second day, a Sunday, proceeded in much the same fashion as in the first day, with lunch at the beach, karaoke singing and others taking a boat ride afterwards or swimming. We heard mass in the resort chapel in the afternoon, before we had dinner in the resort auditorium and to have our dance party. The party was, indeed, the culminating part of our grand golden graduation anniversary. All the participants enjoyed and danced the night away. Of particular note was the skill and art that Rhoti displayed in ballroom dancing! He looked like a dancing instructor, complete with dark long-sleeved shirt and pants. I was watching in awe our high school valedictorian and medical doctor take to the dance floor so beautifully and artistically, especially while dancing with Nila Nayo. The night ended with all of us holding hands in a large circle while singing "If we hold on together", a memorable, nostalgic song by Diana Ross, the most appropriate for the occasion.

MHS '59 luau beach party, Mar. 14, 2009

Rhoti and Nila ballroom dancing, Mar. 15, 2009

I learned that the U.S. group had another reunion in Las Vegas in 2010.

Steps for Life

One day in January 2000, our neighbor and kumpare *Roberto* (*Bert*) *C. de Leon*, invited us to join him and his friends in walking around in the Holy Cross Memorial Park (HCMP), which is just a kilometer away from ACF Homes, our subdivision. He said they walked as soon as the park gate is opened at 6 a.m. and went around the 3.5 km perimeter of the park daily, except Sunday, for exercise. Afterwards they sat and had coffee and pandesal at the chapel.

It was still dark that I could hardly see anyone when Connie and I decided to go to the HCMP the following Saturday at 6 a.m. After parking our car in front of the chapel and walked clockwise or to the left towards the first corner, I was surprised (more like scared in a cemetery) to hear a female voice saying that they have gone ahead and advised us to follow her. We followed her until sunlight came as we walked around the perimeter of the park. We found out later that the woman was *Araceli* (*Cely*) *Gonzales*. We did a complete circuit and stopped at the chapel, where we found Bert and his friends already having coffee. He introduced us to about 15 people who were mostly seniors. We came to know the oldest *Amado* (*Mading*) *Nery*, who was then 78, a retired Quezon City Police sergeant and his wife *Elorda* (*Jessie*) *Serrano Nery*. *Salvador* (*Buddy*) *Rodriguez* was the second oldest and a retired Manila police captain. His brother, Fidel, was also there together with *Andres* (*Andy*) *V. Soriano* and wife Nena. We likewise became friends with *Bonifacio* (*Boni*) *de la Cruz* and wife *Baby*, *Prudencio* (*Pruding*) *Bartolome*, *Ester Bautista*, *Eva Caliwag* and husband *Fred*, *Ligaya* (*Guy*) *Carrasco*, *Amelita* (*Lit*) *San Andres Ramos*, *Corazon* (*Cora*) *Sevilla*, *Maria Fe Canton*, *Wigberto* (*Berto*) *Austria* and wife *Dolores* (*Loleng*), *Eligio* (*Billy*) *Miranda* and wife *Marilyn*, *Josepito* (*Pete*) *Valencia*, *Albino* (*Ino*) *Muyot*, *Felipe* (*Ipe*) *Alcause*, *Celestino* (*Tino*) *Bengcang* and *Ramon* (*Monching*) *Lumaban*. *Ernesto* (*Yoyong*) *B. Rodrigueza* came after his father, Fidel, died. *Michael* (*Mike*) *S. Nery*, *Danilo* (*Jojo*) *L. Sison* and *Larry Tiolengco* also joined the group later. Much later, *Ernesto* (*Nes*) *L. Leaño*, his older sister *Laura L. Mabanag* and her husband, Atty. *Alfredo* (*Fred*) *Mabanag*, also joined the group. Nes was a retired policeman from Makati City. Buddy explained, when I asked, that he had the surname Rodrigueza when he was in elementary school in Bicol but his teacher kept spelling his family name without the letter "a", which became his name from then on. *Margot Casas* also briefly joined the group but left, together with Cely, to dance-exercise with another group. We do not want Billy to visit anyone at home because he usually brings a tape measure, which he uses for his funeral parlor business (Lol).

Billy initially named the group "Sunrise Joggers", then later changed it to "Steps for Life", which I think is more appropriate since we do not jog but walk! Everybody enjoyed our chats and banters after walking and having the usual light breakfast of pandesal with pansit bijon, Queso cheese or Maling meatloaf interchangeably or jointly used as "palaman". Yoyong had always the most frequent and loudest laughs, especially when he teased Berto. We used the name Berto to distinguish him from Bert, who we labeled "Bert Labyo" because he had visual impairment or malabo na ang mata (poor eyesight). He was always seated at least five meters from the group of men because he was the only smoker among us. Buddy also sat away from us because of the smoke from Labyo's cigarette, which Bert smoked almost at five to ten minutes' interval. "Labyo" is a tangential name, in local dialect "disimuladong pangalan", that really means "labo". We had special breakfasts when someone celebrated his/her birthday or wedding anniversary or other occasions like our despedida (Photos below). The celebrator usually brought home-cooked foods or invited the group to Jollibee at the corner of P. de la Cruz St. and Quirino Highway or other restaurants in QC for breakfast or lunch. We occasionally went out of town to have an excursion.

Our despedida breakfast with the Steps for Life walkers on Aug. 18, 2011 before we left for Canada

Our despedida breakfast with the Steps for Life walkers on Aug. 18, 2011 before we left for Canada.

We missed this group so much that we were very happy to find on July 21, 2019 a new group of seniors in Brampton, who are all Filipinos. We met *Conrado* (*Totie*) *E. Pabellano* and wife *Erna*, who were so friendly to us when Connie and I have just finished our lunch and I was having my usual Tim Hortons dark roast coffee at the Bramalea City Centre (BCC). We had a casual chat during which they told us, among other things, that they lived in a condominium unit near the BCC for which they paid a very low monthly amount of less than C$800 with free breakfasts, utilities (electricity, water, and heater/aircon) and garage space, plus unlimited bread every Wednesday. They encouraged us to apply for the seniors housing privilege provided by the Peel Region to its qualified senior citizens. They informed us that when we have spent at least ten years in Canada, we will be qualified to apply for the privilege. Totie advised us to file our application now, in time when we would have reached ten years in the country in 2021. I have applied online and submitted the requirements asked by the Peel Access to Housing or P.A.T.H., which are the proofs of our citizenship and income, besides the form it sent, that I signed. I have photocopied our Canadian passports and our latest income tax declaration and have mailed it to PATH.

Connie's 65th birthday

Connie reached her 65th birthday milestone on August 31, 2009. We prepared a big celebration for her at the Aguirre Resort, about a kilometer from our home in ACF Homes, where we invited for dinner our friends from Steps for Life, our neighbors and my classmates at the Manila High School, and my sister and her children, besides our children and their families. It was a memorable night with a lot of dancing beside the pool, after the sumptuous dinner of a set menu we pre-selected, topped by the crispy and delicious lechon we bought from Elar's that is comparable with Cebu's sauce-less lechon. Our son, Jude, who came

with his wife and two daughters, was able to take many pictures that he shared with me. However, I could not locate the photos now, except for the one at right. I am so happy and grateful to our Lord in Heaven for giving us (her children and grandchildren then, and great grandchildren now) a great wife, mother, grandmother and great grandmother, and a friend of everybody, who is so religious that she reads the Holy Bible before and after going to bed and prays the rosary while we are walking or riding in a bus. I know that God loves her so much and will always take good care of her. Thank God she just celebrated her 75th birthday this year! I am praying that she will celebrate her birthdays for many, many more years to come.

Typhoon Ondoy

Beginning on the evening of Sep. 25, 2009, typhoon "Ondoy" wrought havoc and devastation over Metro Manila, by dumping a 24-hour accumulated rainfall of 455 mm. Many homes were inundated, some up to the second floor, particularly in Marikina City, which is along the Marikina River that swelled above its banks. We did not experience the rain waters going inside our house because it is in the upper sloping part of the hilly terrain in San Bartolome, Novaliches, Quezon City. I have made a large barred hole on our rear concrete fence to allow rain water to flow to a creek located some 500 meters behind our house, beyond our neighboring Greenheights subdivision. Without the hole before, rain water sometimes flooded the interior of our house. Our friends in the opposite slope of the hill in Odelco Subdivision were not as fortunate as us. The Odelco Subdivision is right beside a narrow creek and is about 20 meters below the hill crest at its entrance. *Margot Casas*'s American partner texted me often, giving the status of the water level in their house. His last message said that the flood waters have reached the second floor of their house and they were wading in water on the floor. I knew that he was getting worried about their safety and was just embarrassed to ask for help. The following morning, *Billy Miranda* and *Fe Canton* invited me and Connie to join them to see and help our friends in Odelco. Fe brought some food and dry clothes for those in need. We helped clean the mud that covered the first floor of Margot's house. We saw the one-floor house of *Cely Gonzales* all in shambles. She said she sought refuge on the second floor of the subdivision association building. A neighbor of Margot, who we often saw jogging at the HCMP, told us that she tore the ceiling and roof of her one-floor house to escape the swelling flood, together with her daughter. They stayed on the roof until some people rescued them.

The photo at right is similar to the scenes we saw in Odelco. Four hundred sixty-four (464) people died from the onslaught of typhoon Ondoy. It caused 11B pesos worth of damages to agriculture and infrastructure in the Philippines

My experience with typhoon Ondoy reminded me of the absence of the barangay disaster risk reduction and management council in barangay San Bartolome. I have neither heard nor seen any evidence of its presence in our barangay since we lived here in 1974. I would have welcomed serving in the council as a volunteer adviser.

Immigration to Canada

Connie went to Canada in April 2005 to watch our two grandchildren, who were 11 and 9 years old then, when their parents left for work during week days and to bring and fetch them to and from school. It is illegal in Canada to leave children below 12 years old without an adult at home. When I went to get Connie in December 2005, Mich informed us that she has filed for our immigration to the country through a sponsorship program of the Canadian government. She told me to read and follow the instructions for filling in the application forms she will forward to me through email. Filling in the online application forms was not an easy task. I had to read the instructions carefully because of the language used, which is not the American English that I learned. I managed to complete the application forms, which were more than 10 pages, and sent it to the Canadian Embassy in Makati City through a courier service.

I actually wanted to come to Canada to possibly gain employment because I felt that I can still be productive after I retired in the Philippines, which was coming soon. I knew that there is no age discrimination there in employment.

It was several months or maybe a year later, when I received a letter from the Canadian embassy instructing us to undergo a complete medical examination in its recommended clinic or medical center. The examinations took several trips to Makati City, where parking a car is quite a difficult task, even for paid parking spaces. Our car was towed to the city impoundment garage when I exceeded the time allowed, which was three hours, on the curb of the street where the clinic is located when we first went there for the medical examinations. Connie and I hailed a taxi and asked the driver to bring us to the impoundment garage. He brought us to a place which was the old impoundment garage and asked some people for the right place. We got back our car after paying the fine and towing service fee of Php1,000. That amount was an added expense for the medical examinations that we were undergoing then.

The results of the medication examinations were good for Connie, but it showed that my blood sugar was high. I was referred to Dr. *Augusto D. Litonjua*, an endocrinologist at the Makati Medical Center for further tests. He found that I had Type 2 diabetes, which he said must be treated first before he indorses our application for immigration to Canada. After three or four more appointments with him, we finally received our immigration papers in December 2010! The papers had an expiration date in September 2011. We happily informed Mich about it and she advised us to enjoy the remaining time we had in the Philippines and to be ready for the long haul ahead in Canada. We did just that and we finally landed as permanent residents in Brampton, Ontario, Canada on August 23, 2011.

I knew that some of my former officemates in PAGASA have previously immigrated, too, to Canada, particularly Jun Rellin and *Loreta* (*Lorie*) *Marquez*, the person who provided the information to me to apply for immigration at the Canadian Immigration Consultancy in Pasay City earlier.

Welcome lunch given by ex-PAGASA officemates at *Luis* (*Chito*) *Carbonel*'s home in Green Meadow, Mississauga on Sep. 6, 2011.

Welcome lunch given by ex-PAGASA officemates at *Luis (Chito) Carbonel*'s home in Green Meadow, Mississauga on Sep. 6, 2011.

Luis (Chito) Carbonel organized a welcome reunion lunch for us at his home in Green Meadow, Mississauga on Sep. 6, 2011. Those who attended were: *Emma Carbonel, Felicidad (Fely)* and *Ray Villareal, Janelle Reginaldo Acosta, Edgardo (Ed) Gonzalez, Gilda Cerdeña Borja* and *Alvin R. Ramos*. Chito used to be with the Hydromet. Division while Emma, who is Chito's sister, Fely, Ray, Lorie, Janelle and Ed worked with the TMRDO. Gilda was under the AGSSB while Alvin was the former Legal Officer of the PAGASA.

Connie's first birthday in Canada

Our first Canadian Christmas

      Mich and Roland introduced us to their friends in the GTA when they brought us to the home of husband and wife *Tomas* (*Tom*) *Litimco* and *Corazon* (*Cora*) *Salamat Tiongson* in Georgetown, near Brampton, Ontario on Dec. 26, 2011. We met there *Benjamin* (*Jojo*) *Samson* and his wife *Miriam* (*Yam*). All, except Cora, were once employed by the PLDT in the Philippines. This was followed by many more meetings on birthdays, Christmas, and outings. We were introduced by Jojo and Yam to more ex-PLDT friends like *Larry Magdaluyo Lumbo*, *Efren Marquez*, *Glicerio* (*Olive*) *Oliveros*, *Ismael* (*Bodjie*) *Bulatao*, and *Danilo* (*Danny*) *Tamayo Jr.* We always had fun jamming after lunch or dinner and a few cans of beer at the Samson residence in Toronto since Jojo is good in playing a guitar and a piano, Larry always brought and beat his percussion box while Efren provided the lyrics from his smart phone as they sing mostly Pinoy pop songs. I sometimes joined from the sideline in singing songs I was familiar with. I particularly enjoyed them singing medleys of Pinoy street songs like "Banal na Aso Santong Kabayo" and other similar songs sung by groups when having alcoholic drinks in the Philippines.

First meeting at Litimco residence (L); Jamming in Samson home in Toronto (R).

Typhoon "Yolanda"

On November 8, 2013, typhoon Yolanda (international code name Haiyan) made landfall in the central Philippines bringing strong winds and heavy rains. It was one of the most powerful typhoons ever recorded with sustained winds of 235 kph (147 mph) that caused, together with its heavy rains and resulting landslides, at least 6,600 deaths and massive damages to structures. In its Nov. 16 update, the National Disaster Risk Reduction and Management Council reported that the total cost of damages wrought by Yolanda had reached Php 9.46B and the number of deaths at 6,633.

Hardest hit by Yolanda was Tacloban City in Leyte province, where a storm surge occurred that took away one of the weather observers in the city. Ms. *Salvacion Aveztrus* was the duty observer at the Tacloban PAGASA weather station when Yolanda crossed the island that caused a storm surge. She was with her chief meteorological officer, Mr. *Mario Peñaranda*, at that time. Mario told me later that he managed to climb to the station ceiling when the surge water started to rise, but Salvacion could not follow him up and was swept away by the back surge. She was never found after the typhoon and became a casualty of Yolanda. A new weather station was erected in 2019, still near the airport. I wish the PAGASA officials would name the building after Ms. Aveztrus to commemorate her ultimate sacrifice for the service of the country.

The usual blaming practice of many Filipinos ensued after this major disaster that struck the country. Politicians exchanged charges of misconduct during and after the disastrous event. Even PAGASA was not spared from a vile criticism by a famous tv broadcaster, an ex-congressman from Leyte, who said that its personnel do not know what a storm surge is, that is why one of them died. Other critics said that

Tacloban was not prepared for the storm surge although storm surge risk maps were prepared much earlier by PAGASA, in cooperation with PHIVOLCS and another agency. Although retired then, I was closely following events in my former office. I was so furious when I learned of the tv personality's inane comment about PAGASA's personnel that I urged the president of the Philippine Weathermen Employees Association (PWEA) to demand a public apology from that broadcaster. The fellow apologized later, according to the PWEA president.

Many Filipinos have a short memory span. They quickly forget lessons from the past and commit the same mistakes over and over again. Typhoons Ondoy and Yolanda taught us to prepare for floods when heavy rains start to occur. People living in flood prone areas should have lifeboats or any flotation material ready. If living in a single-story house, provide an access to escape through the ceiling and roof when flood waters rise. Government officials, particularly politicians, should pass on to their successors all important practices and procedures they developed and implemented during their tenure, especially those that benefit the public. The storm surge risk maps for Leyte became just a dead file somewhere.

https://www.abc.net.au/news/specials/typhoon-haiyan-photos-before-after/

Ice Storm in the GTA

A natural phenomenon that I have never heard or learned before was forecast to occur in the Greater Toronto Area (GTA) on December 22, 2013. An ice storm was predicted starting in the evening, that indeed, happened and did a lot of damage to the environment in the

GTA. Curious to find out what the phenomenon is all about, I ventured to walk the following afternoon our dog, Rocky, to the Calvert Park near our home in Tara Park Crescent in Brampton, Ontario. The following pictures depict the effects of the ice storm on all exposed surfaces. Ice coated tree branches and twigs, electrical wires and even our clothes line. As it was already past autumn when the tree leaves have fallen, the ice-coated branches and twigs became heavy and brittle due to subzero temperatures, and many fell because of the weight of the ice. The ice coating was 3 to 4 times the diameter of the twigs/branches (See photo below). Some electrical wires and whole trees also fell, some on parked cars, others on fences. All in all, I thought then that an ice storm was relatively more destructive to the environment as there was no shield from the icing, unlike the winds and rains of tropical storms.

I have to share an embarrassing or funny incident relative to this phenomenon. In the photo below, I tried to step across those fallen trees but it was so wide that my foot slipped after crossing on the icy trail, which caused me to fall on my butt. Luckily, I have a big butt and was unhurt by the fall. A passing man at the far end of the trail saw me sitting upright and yelled if I was alright. I just raised my thumb up to indicate that I was okay.

Climate Change

Our knowledge of climate change began with intense debates among 19th century scientists about whether northern Europe has been covered with ice thousands of years ago. These debates were followed by many more, attributing the recent rise in global temperature mainly due to increased carbon dioxide (CO_2) in the atmosphere as a result of burning fossil fuel.

People, including scientists, often debate about issues in which they have personal or professional interests. Climate change is one of the important issues in the current federal election campaign in Canada. I wish, 'though, that the candidates would foster closer co-existence with nature and other creatures on this planet more than the economic prosperity of the country. Without further contributing to confusion about the issue of climate change, I wish to provide some scientific facts that are relevant to the topic of discussions nowadays.

One, carbon dioxide, methane and other greenhouse gases, are minor components of the atmosphere (0.040862%), as compared to nitrogen (78.08%) and oxygen (20.95%). Two, the age of the Earth is estimated to be 4.54 billion years. It has undergone several glacial and interglacial, called Holocene, periods in the past 10,000 years. The average temperature change ranged from about +5 degrees to about -3 degrees Celsius while the change in concentration of carbon dioxide (from ice cores) averaged about 257 ppmV (parts per million by volume) from 10,000 years ago that rose dramatically over the last 400 years to almost 385 ppmV. Various computer models predict that the Earth's average temperature will rise between 1.8 to 4.0 degrees Celsius during the 21st century. This is still well within the temperature variability in the last 10,000 years.

The Earth's axis moves in a circular motion like a spinning top does. This motion is called precession and its period is about 26,000 years. Precession is caused by the gravitational pull of the Sun and the Moon on the Earth. Due to this, the north celestial pole now points toward the star Polaris. It is quite easy for ordinary people to locate Polaris, to serve as a directional guide at night with a clear or unobstructed sky. It is the star to which the outer lip of the Big Dipper or Ladle points. In 3,000 B.C., the north celestial pole coincided with Thuban, a star in the constellation of Draco. Star Vega in constellation Lyra will be the northern pole star in 14,000 A.D. (http://astrosun2.astro.cornell.edu/ academics/courses/astro201/ earth_ precess. htm). The Sun likewise undergoes a cycle of intense activities on its surface. The cycle that we observe takes about 11 years, which could be the small fluctuations in a much larger cycle that lasts for hundreds or thousands of years. This characteristic is also observable in atmospheric motions like the winds and other natural phenomena. A periodic solar event called a "grand minimum" could overtake the sun, perhaps as soon as 2020 and lasting through 2070, resulting in diminished magnetism, infrequent sunspot production and less ultraviolet (UV) radiation reaching Earth — all bringing a cooler period to the planet that may span 50 years (https://www.livescience.com/61716-sun-cooling-global-warming.html).

The webpage http://www.climatechangefacts.info/ has an extensive discussion of Global Climate Facts. It offers, among others, actions we should and should not take to respond to climate change. I was particularly amused by the suggested action below:

- Should not stop breathing even though it would be one of the most immediate steps to slow CO2 emissions.

Without any basis to offer an analysis of what is actually going on in our environment, I would not dare to issue any statement contrary to or favoring the common notion that the global climate has changed. Suffice it to say that the Sun has a major role in the periodic oscillations or fluctuations of the elements of the atmosphere on Earth, including its observed phenomena.

Man has been self-centered since the beginning of time and has forgotten about his origin. I have been skeptical about scientific explanations of how the Universe evolved from a singularity and expanded into what we now see as the observable Universe. Many people overlook the fact that the Earth and the Universe have been in existence for billions of years and will continue to be there until its Creator decides otherwise. We need to continue praying to God to keep this beautiful planet and Universe He created.

Caribbean Cruise 2014

On January 11 to 19, 2014, we experienced a most memorable series of events in celebrating our 45th wedding anniversary in a cruise in the Caribbean aboard the Celebrity Cruises' Silhouette. It began on our anniversary when we set out to proceed to Fort Lauderdale (FLL), the embarkation port of the cruise. There were no announcements of any changes in our itinerary before we left home but when we checked in at the Pearson airport in Toronto, our harrowing ordeal began!

We learned that our 11:50 a.m. flight to Philadelphia has been delayed, at first, due to the computer bog down since the devastating ice storm before Christmas day last year. There was a later announcement that our flight has been cancelled and we were re-booked to, initially (again!), to 2:30 p.m. flight to Charlotte (CLT), NC with onward connection to FLL. This was later changed to 7:50 p.m. due to further delays. After more delays, we arrived in CLT at 10:20 p.m. but missed the 10:30 p.m. connecting flight to FLL. We were re-booked to the 1:15 p.m. (Jan. 12) flight to FLL and was waitlisted on the 9:50 a.m. flight. We were advised to come at about 8:30 a.m. to see if we can get on the earlier flight, so we decided to stay at the airport to ensure that we will be there early. Our first experience of making the airport our "hotel" was, as expected, not comfortable, to say the least. We tried to catch some sleep sitting on the bench at the check-in area, because the airport security police drove us out of the pre-departure area. I thought I got about an hour of sleep. When my usual night visit to the washroom (bathroom for the Americans, CR for Filipinos) came at about 5:30 a.m., I chanced upon the check-in counter where there were two ground crew waiting for passengers to check in. I asked if we can be accommodated in the first flight to FLL, explaining that we have to be there before 12 noon to catch our cruise ship and showed our boarding pass at 1:15 p.m.

Without saying a word, the person started pounding on his laptop and gave me two boarding passes to FLL, saying "It turns out there were two seats open"! Yehey!

We arrived at the FLL seaport terminal at about 12 noon, very tired, sleepy and hungry but became much alive when we saw our son, Chris, waiting for us at the passenger check-in terminal of the Silhouette! Chris was the ILounge Manager at the Silhouette.

The cruise was to the west Caribbean from Jan. 12 to 19, 2014. The ship "sailed" out after 12 noon from FLL. Chris first checked us in his cabin which is on deck 2 of the 13-deck ship. We were only a few meters from sea level and we can see the waves through the port window. We first had a sight-seeing and picture-taking tour of most of the cruise liner, especially its mess hall, the theater and the casino. In the evening, Chris treated us to dinner at the Disco dining as his additional gift for our wedding anniversary and to welcome us to his cruise liner.

Cruise passengers who opted to join the land tours were transported ashore, in case the ship cannot dock, through a small boat called "tender". Connie and I decided to join four land tours.

Our first stop was on Jan. 14 in Cozumel, Mexico, where we saw a small ruin of a Mayan pyramid, now covered by small trees, and a small Mayan village with men whose entire body is covered with decorative paints. There was nothing more to see in Cozumel, except the town plaza which was similar to an old town in the Philippines. Our second stop was on Jan. 15 at the Grand Cayman Island, where our tour guide promised to bring us to Hell and back. He did as he promised after showing us why the town was named Hell, when we visited a place with an eerie and desolate looking landscape composed of sharp gray colored rocks that are a result of the solution of the rocks due to acid rain and its consumption by carbonate-loving organisms (See top photo on page 147). Even Hell has a church with the sign "Christ promised to save you from hell" painted on its roof and a post office next to the church (See lower photo on page 147).

First formal night dinner at Qsine fine dining restaurant.

We next took the mini-submarine tour in Grand Cayman Island. The "submarine" was actually a boat, with its submerged hull provided with seats on both sides where there are large enough windows to see the sea floor. The tour was about 30 minutes, which took us around a port where pieces of the wreckage of one of Columbus' ships lie on the sea floor. I was completely satisfied with the entertaining explanations of our Jamaican tour guide that I remarked to him before alighting from the submarine that he is like his compatriot Usain Bolt. When he asked me why, I said to him that they both make their country proud. He smiled and shook my hand in gratitude.

The Silhouette's next stop was at Falmouth, Jamaica. It was raining when the ship docked at the port in the morning of Jan. 16. Connie and I waited at the Oceanview Café of the ship until the rains stopped to walk around the port since we did not join a land tour here. We bought though a colorful long dress for our great granddaughter, Cianna, at an open market at the port.

The last stop of the cruise ship was at Labadee, Haiti on Jan. 17. Labadee is a private resort of Celebrity Cruises in Haiti, where our ship had a pier to dock in. It is on the southern coast of the island where a magnitude 7 earthquake hit the island near Port-Au-Prince on Jan. 12, 2010. The death toll in this quake varies from 200,000 to 300,000 (CNN). Its main attraction is the Columbus' Cove Beach, which has white sands all around. We joined a walking tour around the cove, where the tour guide, Mr. *Lamy Docleur*, gave us very important and quite interesting historical information about Haiti, including a tree called "Neem" when we walked under some of it. He said that the tree has many medicinal properties, like for blood pressure control and cure for cancer. He added that it is very expensive to buy its leaves and it is prohibited to sell it overseas.

We had our second formal night dinner at the Grand Cuvee fine dining room on deck 5 of the Silhouette. I had two pieces of over-roasted oysters while Connie and Chris had six pieces of escargot or kuhol with coconut milk, as appetizers. They both had salad while I had a tomato soup next. We all had a lobster soaked in butter and lemon as our entrée. For desserts, Connie and I each had an Alaska ice cream cake and shared a cheesecake while Chris also had an Alaska ice cream cake and a leche flan with burnt sugar icing.

In spite of the ordeal we had to hurdle to get to FLL, we had a wonderful experience in celebrating our 45th wedding anniversary, much worth the great difficulty we went through and the costs that our daughter and son incurred to get us there. We hope to do it again soon!

With our son Chris in Labadee, Haiti.

New Canadian and dual citizens

Connie and I decided to apply for Canadian citizenship in 2014, after having experienced the excellent government system in the country, as compared to my country of birth. As permanent residents, we were already entitled to all the benefits extended by the government, except the right to vote. Most significant for us is the Ontario Health Insurance Plan (OHIP), which is provided free of charge to all Canadian citizens and permanent residents in Ontario. OHIP covers all medical services, including professional fees, most prescription medicines and laboratory services, except dental services, eyeglasses and hearing aids. I do not pay my family physician for his services and for my maintenance meds, which are home delivered every other week. I have had a prostate surgery, cataract operation on both eyes, and two colonoscopy surgeries, all without being charged a single cent. We could not have afforded all these services in the Philippines with our meager pensions, which are enough only for our basic needs including my prescription medicines for diabetes and for cholesterol, uric acid, and blood pressure control.

In addition, the Ontario government provides convenient transportation systems to its citizens, most significantly to persons with disability (PWDs) and the elderlies. The government operates the only transit and train system in the province. It should be stressed that all timing systems in the country are all synchronized and bus schedules are closely adhered to by the operators (persons who operate the vehicle; they are not called drivers). This means that all passengers should observe the schedules if they wish to be on time for their activities. Buses serve all major roads in Brampton, with stops at convenient places. The buses kneel for easier embarkation/disembarkation by PWDs/seniors and have a platform that unfolds when a passenger in wheelchair wishes to ride or disembark. There are spaces on the bus reserved for these passengers, as well as for pregnant women and baby carriage, near the driver's area. There is no overhead pedestrian bridge and all pedestrian walks have a ramp, including all buildings. Seniors pay only C$1 cash for a two-hour trip anywhere in Brampton and its southern neighboring city of Mississauga.

I filed our application for citizenship on Nov. 5, 2014, a day before I underwent a prostate surgery at the Brampton Civic Hospital on Bovaird Drive. It took about 10 months when we received a letter from Service Canada, advising us to proceed to the Scarborough Town Centre on Oct. 6, 2015 for our oath of citizenship. We took our oath before Judge *Marian Sami* (Top photo next page).

I was not aware that, upon taking the oath of Canadian citizenship, Connie and I lost our Filipino citizenship. I thought that, since our passports were still valid up to 2017, we were still Filipino citizens. When I found out that we were not, I made an online search to find out how to be a dual citizen. Connie and I went to the Philippine Consulate offices on Eglinton Ave. in Toronto to apply for re-acquisition of our Filipino citizenship. We were so pleased that Deputy Consul General (DCG) *Bernadette Therese* (*Bernie*) *C. Fernandez* personally received us in the conference room and made us fill in the application forms. She then instructed us to pay the corresponding application fee and advised us to return upon receipt of a letter for our oath taking. It did not take long before I received the letter and we took our oath of Filipino citizenship before her in the conference room. Bottom photo below shows DCG Bernie Fernandez administering the oath of citizenship to some Filipinos at the Philippine Consulate in Toronto. We were extremely happy for DCG Bernie Fernandez service to us that I thought that she will become a full-fledge Consul General in the near future. She is an ideal Filipino government servant and deserves the highest respect and reward for her outstanding service to her fellow Filipinos abroad.

We became Filipino citizens again in August 2016 and became dual citizens. Someone jokingly called us later triple citizens, i.e. Canadian, Filipino and senior citizen all at the same time!

Marians in the GTA

We became aware that Mich had become connected with some of her high school classmates in the Greater Toronto Area (GTA) after we arrived as permanent residents in Canada on Aug. 23, 2011. She told us that *Perpetua Rosario (Rosepet) Tan Dimatulac*, her husband *Raymond* and children *Cedric* and *Carlos*, stayed with them in their apartment in Scarborough in 2004 before Mich and her family moved to Brampton in 2009. Rosepet was Mich's classmate at St. Mary's College in Quezon City (SMCQC) from where they graduated in 1987. We likewise learned that Mich had found her other classmates like *Marla Atienza Feliciano*, who was their class valedictorian, *Clarinda (Claire) Concha Andres*, *Elizabeth (Beth) Estacio Crisostomo*, *Regina (Reggy) Calaguas*, *Perla (Perlie) M. Suarez* and *Jhoy Sanchez Labudahon*.

Shortly after we arrived in Canada, when Mich and her family have moved to Brampton, my former officemate *Loreta (Lorie) Marquez*, called to arrange an appointment with us at home one day. She said she wanted to introduce her friend to us for a business proposal. When the two arrived, Lorie introduced us to *Antonia (Jingle) Vera Canaon*, who is a medical doctor in the Philippines but was now working as an assistant in a hospital in the York region. She was pleasantly surprised to see Mich as both recognized each other as classmates in SMCQC. After the happy reunion with Mich, Jingle explained to us her sideline work in selling personal life insurance. We promised to study her proposal and invited her to return when she is free. Jingle became an addition to the GTA Marians since then, although she attends its social activities irregularly due to her work and greater distance from most her classmates' residences.

The GTA Marians had frequent family activities since we arrived. They invited us and their parents to birthday celebrations and picnics during the summer school vacation. Below are examples of their close bond as high school batchmates. The left photo below is one of the few occasions where all nine of them were present. Seven of the Marians are married, except Perlie and Reggy, who are both single. Marla is married to *Jerome Feliciano*, Jhoy to *Magdy Samwaeil Ramzy*, Claire to *Joel Reyes Andres*, Beth to *Henry Andrada Crisostomo*, and Rosepet to *Raymond (Dindo) Dimatulac*. Jingle is married to *Antonio Canaon*, who is also a medical doctor like her. The Felicianos have their daughters Jelene and Juliane, while Jhoy and Magdy have Jhanlyane, Maryam and Margo as their kids. The Andreses have Joaquin and Mary while the Crisostomos have Elaine, Gian and Lia. The Dimatulacs have their sons Cedric and Carlos. The children are almost always with their parents whenever we had special occasions, except the Canaons, who rarely attend such occasions. These youngsters are all properly brought up by their parents to be respectful by keeping the Filipino tradition of touching the hand of their elders on their forehead when they meet them. They are so smart and talented that I foresee them to be successful in their respective future careers. Connie and I love them and their families so much, like they were our own children and family.

Connie and I have become happier since we have become part of the GTA Marians' family. We are grateful that we have met *Lilia* (*Lily*) *Atienza, Pacifico* (*Peping*) and *Emilia* (*Mia*) *Feliciano, Remigio* (*Remy*) *Y. Tan, Evangelista* (*Ely*) *Estacio*, and *Rodrigo* (*Rod*) *Calaguas* and wife *Norma* (*Ming*). They are the parent/s of Marla and Jerome, Rosepet, Beth, and Reggy, respectively. We also met briefly *Nestor* and *JR*, father of Marla and Beth, respectively, and *Josie* who is the mother of Dindo. We cherish the moments when we are together, especially when we spend videoke singing with Lily, who is quite good as she often gets a 100 rating!

Jingle, Jhoy, Claire, Rosepet, Remy (Rosepet's papa), Reggy, Mich, Beth, Perlie and Marla on the 80th birthday of Remy.

Another addition to our circle of friends and families came on June 5, 2016 when the Lagura family arrived from the Philippines. *Leila* (*Lei*) *Carpio Lagura* joined our household with husband *Randie*, daughter *Bianca Cyril* and son *Reeno Kurt*. Lei is a younger sister of our son-in-law Roland. She and Randie are our wedding godchildren. Lei was a registered nurse in the Philippines. She is now working in a food manufacturing plant in Brampton. Randie once worked as a chef aboard the same cruise ship as our son Chris. He is a quiet and unassuming person. He now works as a chef with our grandson Miro as a retail assistant in a large grocery chain along Bovaird Dr. also in Brampton. Bianca is now in Grade 12 at the Cardinal Leger Secondary School while Kurt is in Grade 4 at the St. Anne Catholic School. Both schools are in Brampton with St. Anne just five minutes' walk from our house. Bianca is a voracious reader and got a grade of 99% in English last school year (2017-2018). I said to her that she is much better than I in that subject. She dreams to be a surgeon someday. I am confident that she will achieve her dream. Kurt is still a growing boy but he is also smart like his sister. Connie and I love this family, too.

Marian family children Carlos, Joaquin, Miro, Cedric, Margo, Jelene, Mary, Rochele, Maryam, Lia and Jhanlyane. Juliane is not in the picture.

The Lagura family on Kurt's 7th birthday celebration in Aug. 2017

On Christmas day, 2018.

Third family generation career movements

Vernie, our oldest grandchild, completed her B.S. in Hotel and Restaurant Management at the Philippine Women's University in Quezon

City in 2009. We were not informed of her graduation and we do not have any picture of her important life milestone. We are so proud of her for her independence and resourcefulness that she does not wish to impose herself upon anybody. Vernie is now the cashier at the Orange Bucket near Rocka Ville in Plaridel City, Bulacan. Next is Rochele, who completed a B.S. Nursing at the University of Ontario Institute of Technology in Oshawa, Ontario in June 2017. She is now a Registered Nurse and is a Cosmetic Nurse Injector at the Skin Vitality Medical Clinic in Toronto since January 2019. Cha followed suit when he graduated with a B.S. in Information Technology at the Centro Escolar University in Malolos, Bulacan on April 16, 2018. He is now working as a Job Order at the Human Resources Management Section of the PAGASA. Lastly, Miro graduated with a B.S. in Medical Science at the York University in Toronto on June 16, 2019. He is still working as a retail assistant at the Fortinos on Bovaird Drive in Brampton. We are confident that he will soon find a better and more suitable job for his qualifications.

Our 50th and Mich-Roland's 25th Wedding Anniversary

Connie and I had our 50th wedding anniversary on January 11, 2019 while Mich and Roland had their 25th wedding anniversary on January 8, 2019. Mich arranged that we concelebrate our anniversaries on Jan. 12, starting with our renewal of vows at the St. Anne Catholic Church on Vodden St., Brampton. We then proceeded to the Lunchbox Café and Catering on Kennedy Rd. in Brampton for lunch of various Filipino dishes. Mich invited *Neil Huab* and wife *Janette* to provide the musical entertainment. All the GTA Marians, their husbands and children came, including some of my friends. I was so elated that *Chito Carbonel, Alvin Ramos, Alice Rafol* and *Lyzel Siao* joined us in our moment of happiness at the Lunchbox Café. Chito and Alvin are my former officemates, Alice is my MHS co-alumnus, while Lyzel is a daughter of my best friend and MHS batchmate, Star Siao. Alice came with her husband, *Gil Rafol.*

With Rev. Fr. Jibin Joy, who officiated our renewal of vows at St. Anne Catholic Church.

Connie and Bernie pose at home before going to St. Anne Catholic Church.

I was requested to sing two songs at the start of the program to serenade Connie and I chose "Hanggang" by Wency Cornejo which is her favorite song. I also sang "Through the Years" by Kenny Rogers which is my favorite one. It is regretful that I was unable to take pictures of the most joyous parts of the occasion, which are the "Name that song" game led by Neil and the dance presentations. I only took videos because all the people were in motion. Anyway, the videos are in my timeline on FaceBook for those interested to watch them. It is quite fortunate that Alice was able to take some snapshots of the momentous occasion, which are shown below and the following page.

Connie and Bernie prepare to cut their wedding anniversary cake.

Lyzel Siao, Gil and Alice Rafol.

Brampton Filipino Seniors Club

Connie and I used to go out every Sunday since the end of winter 2019 to go to the malls, either at the Bramalea City Centre (BCC) in Brampton or at the Square One mall in the Mississauga city centre. We went there to catch some fresh air and see sights other than the four walls of our home. As I mentioned earlier, we met *Conrado (Totie) Pabellano* and his wife *Erna*, together with *Aurora (Au) Espejo Caume* at the BCC on July 21, 2019. Totie is the vice president of the Brampton Filipino Seniors Club (BFSC) while Au is its treasurer. Even if we were not members of the club yet, they invited us to their annual summer picnic at the Chinguacousy Park near the BCC on Aug. 10.

The BFSC held its monthly general membership meeting (Photo below), chaired by its president *Peter de la Cruz*, at the Knightsbridge Community and Senior Citizens Centre, Brampton on Sept. 21, 2019. Among other items of the agenda was honouring the birthday celebrants for the months of August and September and welcoming the new members. Connie and Erna joined the birthday celebrants while Connie and I and *Lydia Bonzon* were warmly welcomed by the Club as its new members, led by its president, vice president and assistant vice president *Conrado (Radd) Maramag*. I met *Chary Abuyuan*, the youngest sister of my former office colleague Ching Belmonte, during the general membership meeting. I was so impressed at Au's presentation of her monthly financial report that I felt comfortable being a new member of the BFSC.

The old and new members. Photos courtesy of Tim Evangelista.

Bucket and "pre-departure" check lists

There are still a few items in my bucket list that I hope to check out soon. One is to see and capture in my video and/or still camera the Aurora Borealis, the so-called "northern lights". A second place I wish to visit is the Christ the Redeemer statue in Rio de Janeiro, Brazil. The third is the great pyramids of Giza, Egypt. Fulfilling the second and third destinations in my bucket list would result to my having been to six of the seven continents of the world, namely; North America, South America, Europe, Asia, Africa and Australia. The only continent left is Antarctica. I just found out that there are cruises, tours and trips to the coldest continent on Earth via the southern tip of South America, either Argentina or Chile. I hope Connie and I will still be healthy enough and that our financial status would allow us to go to the remaining places in my bucket list.

I like the expression "Live for today and don't worry 'bout tomorrow, for tomorrow will worry about itself". However, we care so much for our loved ones that we do not wish to burden them when the inevitable comes. Connie and I have, therefore, prepared for our final "destination" by getting two plots in the Holy Cross Memorial Park in Novaliches, Quezon City, as early as in the 1980's. This year (2019), Mich got us a funeral plan and we have signed a Will, witnessed by Leila C. Lagura and Roland Miro S. Carpio. The Will just needs to be notarized to become a legal document.

I will continue to be of service to my loved ones and to the world in whatever capacity I can. I owe it to God that I had a wonderful life on Earth and I wish to be His messenger of love toward His children. I wish that peace and harmony will reign all throughout the world. Let us "Imagine" and "Bless the Beasts and the Children", as John Lennon and Karen Carpenter gently urged us in their song, respectively!

I will never forget the following scenes where my granddaughter, Blessie Marie (Photo below), and great granddaughter, Cianna Haylee (Photos next page), are shown in tears as they long for their parents who were leaving for work. Both were about the same age when these pictures were taken. Blessie lived with her mother separately from our son Jude, who was then bringing his other daughter to school every morning before going to work. Blessie misses her father so much that she did not want him to leave her. Only a person with a heart of stone, or "pusong bato" in our native language, will not feel sad in seeing a child crying for his/her parent.

Building Monuments in the Sky   *Bernardo Soriano Jr.*

Cianna also did not want her mother to leave her for work. Vernie told her that she must work so that she can buy her "dede" (milk). Cianna tearfully said "Di na ako dede, wag ka work" (I will no longer take milk, do not work). I am so thankful that our great grandchild, at a tender age of two years, had already a sense of self-sacrifice for her love of her mother. Thank God for our granddaughters' love for their parents. Our two granddaughters and others like them inspire me to continue to be of service to the world!

I solemnly hope and pray that God will give me a healthy mind and body to achieve my further aspirations until He calls me to listen to His answers to my questions "Was I brave and strong and true? Did I fill the world with love my whole life through?"

I fervently hope that God will answer each question with "Yes, my son".